Hypermodernity and Visuality

Critical Perspectives on Theory, Culture and Politics

Critical Perspectives on Theory, Culture and Politics is an interdisciplinary series, developed in partnership with the Centre for Critical and Cultural Theory, which is based in the School of English, Communication and Philosophy at Cardiff University, UK. The series focuses on innovative research produced at the interface between critical theory and cultural studies. In recent years much work in cultural studies has increasingly moved away from directly critical-theoretical concerns. One of the aims of this series is to foster a renewed dialogue between cultural studies and critical and cultural theory in its rich, multiple dimensions.

Titles in the series:
Culture Control Critique: Allegories of Reading the Present, Frida Beckman
Prometheanism: Technology, Digital Culture and Human Obsolescence, Günther Anders and Christopher John Müller, translated by Christopher John Müller
Creole in the Archive: Imagery, Presence and the Location of the Caribbean Figure, Roshini Kempadoo
The Attention Economy: Labour, Time, and Power in Cognitive Capitalism, Claudio Celis
Performative Contradiction and the Romanian Revolution, Jolan Bogdan
Chinese Subjectivities and the Beijing Olympics, Gladys Pak Lei Chong
The Extreme in Contemporary Culture: States of Vulnerability, Pramod K. Nayar
Superpositions: Laruelle and the Humanities, edited by Rocco Gangle and Julius Greve
Credo Credit Crisis: Speculations on Faith and Money, edited by Laurent Milesi, Christopher John Müller and Aidan Tynan
Materialities of Sex in a Time of HIV: The Promise of Vaginal Microbicides, Annette-Carina van der Zaag
From Shared Life to Co-Resistance in Historic Palestine, Marcelo Svirsky and Ronnen Ben-Arie
Affective Connections: Towards a New Materialist Politics of Sympathy, Dorota Golańska
Homemaking: Radical Nostalgia and the Construction of a South Asian Diaspora, Anindya Raychaudhuri
Hypermodernity and Visuality, by Peter R. Sedgwick
Partitions and Their Afterlives: Violence, Memories, Living, edited by Radhika Mohanram and Anindya Raychaudhuri (forthcoming)
Contested Borders: Queer Politics and Cultural Translation in Contemporary Francophone Writing from the Maghreb, William J. Spurlin (forthcoming)

Hypermodernity and Visuality

Peter R. Sedgwick

Poems by
Damian Walford Davies

ROWMAN &
LITTLEFIELD
INTERNATIONAL
London • New York

Published by Rowman & Littlefield International Ltd
6 Tinworth Street, London SE11 5AL
www.rowmaninternational.com

Rowman & Littlefield International Ltd.is an affiliate of Rowman & Littlefield
4501 Forbes Boulevard, Suite 200, Lanham, Maryland 20706, USA
With additional offices in Boulder, New York, Toronto (Canada), and Plymouth (UK)
www.rowman.com

British Library Cataloguing in Publication Information Available
A catalogue record for this book is available from the British Library

ISBN: HB 978-1-78660-490-3

Library of Congress Cataloging-in-Publication Data

Names: Sedgwick, Peter R., author. | Davies, Damian Walford, writer of supplementary textual content.
Title: Hypermodernity and visuality / by Peter R. Sedgwick ; poems by Damian Walford Davies.
Description: London : Rowman & Littlefield International, [2019] | Series: Critical perspectives on theory, culture and politics | Includes bibliographical references and index.
Identifiers: LCCN 2018059078 (print) | LCCN 2018058277 (ebook) | ISBN 9781786604903 (hbk.) | ISBN 9781786604910 (Electronic)
Subjects: LCSH: Technology—Social aspects. | Technological Innovations—Philosophy.
Classification: LCC T14.5 .S396 2019 (ebook) | LCC T14.5 (print) | DDC 303.48/3—dc23
LC record available at https://lccn.loc.gov/2018059078

Er cof am Bryan Martin Davies
Prifardd

Contents

Acknowledgements

My thanks to my wife, Siân, for her invaluable advice on many parts of this text—and for putting up with someone too often engrossed in writing. Thanks also to Damian Walford Davies for his contribution of poetry. Gratitude is owed also to Andrew Edgar for his, as always, invaluable and astonishing breadth and depth of knowledge. Finally, sincere thanks are due to Christopher Müller for his insightful comments on an earlier draft of the text. Grateful acknowledgement is also due to the Musée du Louvre and Musée d'Orsay for permission to use photographs that were taken in their galleries.

Preface

This book engages with facets of 'hypermodern' visuality. It considers the archaic elements that abide within the unsettlingly modern gaze. The discussions offered take their lead from two directions. The first is allied closely to the spirit of Max Horkheimer and Theodor Adorno's *Dialectic of Enlightenment*.[1] The second is the 'creative-critical' approach that has been developed in recent years by practitioners of literary theory and creative writing.

In relation to the first of these, it should be said that what follows does not engage directly or at length with Horkheimer and Adorno. The discussion instead turns to other writers (not least, Virilio, Nietzsche, Heidegger, and Bauman) in order to initiate its own point of entry into hypermodern visuality, the historical modernity out of which it is articulated and the archaic features which are continually reconstituted within it. What is emulated, though, is the general approach Horkheimer and Adorno develop in *Dialectic of Enlightenment*, which offers an inexorable charting of modern reason's self-undoing under the banner of 'Enlightenment'. According to this view, reason recoils into a means-ends form of thought which, rather than being an 'objective' mode of engagement, becomes a malleable tool amenable to serving powerful social interests. In consequence, instead of generating an ethos for cultivating critical reflection on the value of society's stipulated goals, Enlightenment rationality renders itself the obedient servant of a hegemonic version of 'method'. In the modern era, this hegemony takes the instrumental-calculative attitude of the 'hard' sciences (positivism, in all its forms) to be paradigmatic of 'reality'. For Horkheimer and Adorno, the consequences of this condition erupt decisively with the barbarism of the 1940s and its deployment of instruments of annihilation on a grand scale in total war. Modernity, red in bullet and bomb, reveals its purportedly 'enlightened' character to contain a contradictory autochthonous violence which hovers like an anguished and accusing question over the facade of benign progressiveness with which it advertised itself to the world.

Horkheimer and Adorno's critical engagement leads them to identify the archaic roots of this ruthless instrumentalism in ancient Greek culture. These origins can be detected in the ancient myth of the *Odyssey*, whose eponymous, proto-bourgeois hero deploys a sly and calculating reason to overcome the obstacles that stand in his way.[2] The disturbing quality alluded to above which haunts any contemporary eruption of archaism is, in other words, due in no small part to the fact that although the archaic seems to hark back to something we feel ought to be long gone and so 'behind' us, there remains, as Jean Baudrillard has noted, something 'remarkably modern' and forcefully present about it.[3] Old Odysseus is closer to us moderns than we might like to think. Baudrillard notes that when, on 14 February 1989, Ayatollah Khomeini condemned Salman Rushdie for writing *The Satanic Verses*, he invoked an archaic language of 'Evil' which a horrified, liberal West, with its Enlightenment-inspired faith in a conception of the 'Good' as something pursued through the instrumentally efficient management of material resources, seemed powerless to resist.[4] The disruptive power of a pronouncement like Khomeini's springs from the manner in which its archaic terms of articulation can exploit Western media space and sensibility. This last point suggests a kind of archaic appeal at work in the Western milieu which persists alongside the more self-conscious archaism of a Khomeini—perhaps as a precondition of the latter's ability to exert influence. The archaic is 'modern' in the sense that it abides within the contemporary itself, both from the outset and continually, as a constitutive but also potentially disruptive condition of Western sociality. Above all, the archaic resides in the midst of our dominant practices. How it might do so is explored in the introduction to this volume. For the moment, one can say that, understood in this way, the archaic can be grasped as something implicitly present in the very midst of the contemporary cultural environment in all its 'enlightened', technically efficient modernity—not least, in the most intense and accelerated 'hypermodern' form of this modernity.

It is this presence which the following chapters seek to elucidate. They do so by way of a consideration of hypermodern visuality in particular. The visual is replete with elements that are by turns, and sometimes at once, ancient and modern and hypermodern. We live in the era of the rapidly recorded, rapidly transmittable, unstable, and ubiquitous mechanical image. To explore the hypermodern in the visual register is therefore timely. To consider the contemporary and the visual in relation to the persistence of the archaic, however, is to seek to puncture that sense of timeliness by turning it against itself. One can thereby question the positivistic version of the Enlightenment narrative which allies technical advancement with progress by making visible those ancient aspects of visuality which have a tendency to get concealed by this narrative even as they exert an important influence over it.

A rejection of positivism underlies the creative-critical mode of engagement with hypermodern visuality that is developed here. This approach shares something of the qualities that Adorno attributes to the essay form. It offers a mode of thought that is open-ended. One does not begin, as the positivist does, by approaching one's subject matter armed with a rigid 'method' which settles in advance what is to count as valid or invalid. The positivistic methodological attitude tends to consider one mode of expression (usually a 'logical' one grounded in an empiricist outlook) as being paradigmatic of truth-talk—all else is 'meaninglessness'. Against this, the creative-critical attitude cultivates a 'methodically unmethodical'[5] approach, to adopt Adorno's characterization of the essay form. Following Adorno's conception of the essay, this approach seeks to open out a space of enquiry in which questions show themselves rather than enclosing its objects in an all-knowing gaze. One does not cleave to *a* method; one does not believe one knows all the rules in advance; one does not believe that enforcing such rules necessarily leads to the best of all possible outcomes. It is a matter, rather, of multiplying and intermingling genres of discourse.

In the present text, this intermingling is sought in combinations of images and different varieties of textual reflection (poetry, prose, critical) in order to explore the visual in its hypermodern register. It does so with more than one authorial voice. This is the case not merely with regard to the fact that the poems are written by Damian Walford Davies rather than Peter R. Sedgwick. A multiplicity of voices and perspectives is implicit in the use of images, prose texts, and critical texts and in the dialogue with which the volume concludes. The images presented here are, in this respect, not mere 'illustrations' of 'things'. They are offered as visual 'aphorisms' (and sometimes interjections) to be responded to and interpreted. The written material that accompanies these images seeks, in different ways, to do just this. In this regard, the current volume takes seriously László Moholy-Nagy's contention, made in 1928, that 'the illiterate of the future will be [. . .] the person who cannot photograph'.[6] That is, 'literacy' demands sensitivity to the ambiguities of the visual, to the problematic nature of 'representation'. As with others who have embraced the critical-creative approach, so here, too, many precursors are called upon, e.g. the Nietzschean aphorism, Walter Benjamin's notion of the 'thought-image', even the Platonic dialogue. Some are not even mentioned but remain implicit. Not everything needs to be 'made explicit', after all—indeed, the very notion that 'everything' could be rendered entirely 'transparent' is something that the ensuing discussions repudiate.

The following text is composed of an extended introduction followed by two parts. The introduction, in accordance with the rejection of positivism signalled above, begins by addressing the question of the visual as an essentially cultural condition. More specifically, it seeks to lay out the terrain of the discussion of the visual by endorsing Heidegger's conception of what makes a 'world'. The notion of 'world' is then articulated in terms of the concept of 'practice'. The case is put that visuality, the realm of the seen, is not to be comprehended in terms of the relation between such fictions as 'empirical stimuli' and 'intellectual concepts' which simply 'represent' them (a crude empiricist view) but rather in terms of the historicity of practices which permeates any cultural world. The chapter then elaborates seeing in its relation to modernity and the archaic. Initially, this is broached by way of Nietzsche's account of the relation between (enduring) practices and (temporary and shifting) purposive meanings. The aim here is to show how the historicity of ancient inherited domains of practices persists in and pervades the spaces of meaning which make up the 'present' as an essential condition of its possibility. Attention is directed to the ways in which modernity is both the simultaneous retention and disruption of traditional articulations of visual cultural space. This is captured well by, amongst other things, Baudelaire's famous dual formation of the modern. The account of the hypermodern that is subsequently offered is articulated out of this ambivalent modern context. After casting a critical eye over some claims made by Gilles Lipovetsky, the discussion endorses Zygmunt Bauman's contention that hypermodernity is 'liquid modernity', i.e. the overturning of the boundaries of tradition by social-technological cultural forms which aim to cultivate a limitless transparency. This, however, it is argued, occurs alongside the persistence of more opaque archaic and outmoded elements that remain intrinsic to the hypermodern.

The extended essay which forms part 1 elaborates this persistence by foregrounding Paul Virilio's account of the bunker and the military visual logic of defensive enclosure which founds the city structure and Western culture generally. Virilio persuasively argues that this defensive visual practice is the archaic basis out of which the accelerating 'dromological' modern metropolitan cultural world, like its hypermodern variant, is articulated. The chapter explores this articulation in a number of registers. These include the relentless pursuit of speed which legitimizes faith in pure transparency; the defensive sensibility that renders the contemporary world prey to terror; the hypermodern transformation of city space into a domain of total surveillance; the city's return to the condition of the defensive bunker from which it sprang; and the role of that most paradigmatic of contemporary devices, the camera, in the ancient logic of enclosure which conditions contemporary visuality.

The concluding discussion of part 1 considers critically the forming and shaping of the self in relation to the visual. The engagement offered seeks to go beyond Virilio's metaphysical conception of human identity even as it cleaves to his essential insights. Against Virilio's metaphysical view (which is grounded in his Christian faith), it is held that the modern self can be regarded as being historically fashioned by ancient practices of seeing which emerge with the ancient founding of the city. A radical reinterpretation of Nietzsche's *The Birth of Tragedy* is then developed in order to argue that these practices can be linked to three versions of visuality: (1) an enclosing Apollonian visuality characterized by a defensiveness rooted in surveillance—this visuality forms the public-civic space of custom which cultivates the self; (2) an unruly Dionysian embodied over-spilling visuality which the Apollonian needs and seeks to contain and shape in order to found and maintain civic life; and (3) a subsequent elaboration of Apollonian defensiveness into a paranoid Socratic visuality which, in its desire to enclose the world in the objectifying truth of an all-seeing gaze according to which reason and light are one, both denigrates its own Apollonian origins as mere 'custom' and rejects the embodied excesses of the Dionysian as irrational, so placing itself above them. The logic of contemporary hypermodernization, it is argued, turns on the struggle between these archaic but still present Apollonian, Dionysian, and Socratic visualities, with the latter, in the guise of pure, rational 'method', forever seeking to enclose and overpower the other two. Thus ancient Socratic visuality, in secularized form, lies at the heart of the controlling, positivistic, methodological desire which has come to dominate in the late and liquid modern era.

The discussions presented in part 2 of the volume offer 'methodical unmethodical' elaborations of the theme of Socratic enclosure and its relation to the hypermodern in a variety of registers.

Part 2 chapter 1 begins with an image of people standing, mobile devices in hand, before the *Mona Lisa* in the Musée du Louvre, Paris. The ensuing poem, prose reflection, and five individual, partly self-contained sections then explore the hypermodern electronic communications device, the screen and rippling archaic presences that can be detected lingering in close proximity to them. Central to the discussion is the question of who 'we' are. Each engagement is concerned in its own way with how the screen frames, encloses, and so constitutes contemporary self-understanding and with charting how archaic elements in the midst of this enclosure serve to disrupt it.

Part 2 chapter 2, which commences by returning the eye to Paris and the Musée d'Orsay, considers the positivistic logic of calculation in its visual-observational register. The overt target is the positivistic practice of 'empirical aesthetics'. The discussion ponders such matters as the relation between familiarity and equipment; the possibility of seeing simply in order to see; the implications of the transformation of categories of thought into visual 'information' by method; the notion of instrumentalist visuality as colonialism; the enclosure of gallery and the artwork; modernity and the question of uselessness; and the presence in gallery space of the Dionysian, Apollonian, and Socratic visualities outlined in part 1.

In chapter 3, the text takes on a more personal tone, embracing (. . . in a perhaps disingenuous manner? It is hard to say . . .) a mythological innocence so as to consider how the digital image returns us to a 'pre-photographic' ethos. On the path to this consideration, Barthes and the relation between the photographic image and death are touched upon, as is the condition of the 'analogue innocence' of film and the possibility the digital image might be read as an invitation to engage with an ethics of the visual.

The concern of chapter 4 is the ancient tragedy of Oedipus, and its principle of organization is the relation of the Oedipal to the Apollonian, Dionysian, and Socratic modes of visuality. This link is broached via a return to Virilio's theme of speed (dromology). Aspects of the thought of Freud and Nietzsche are engaged with quite closely, but the central argument is that the play *Oedipus Tyrannus* can be read as exploring how a combination of the Apollonian, visual logic of enclosure and the Socratic, rationalist desire to render the world amenable to the absolute control of human will undermine themselves. Faced with danger, Oedipus places his faith in the quick-witted speed of a cunning foresight capable of encircling, enclosing, and so transcending the earth. To achieve this transcendence would be to insulate oneself from the threat of the unforeseen, i.e. of suffering. However, nobody can see the future that awaits them. As Oedipus flies headlong toward his dream of domination through transparency, he forgets the concealed, embodied Dionysian element which is also part of him. This element explodes out of him as blind Fate. Fate cannot be seen in advance; nor can it be outrun by predictive strategies that place their faith in an all-seeing, rational transparency. Oedipus's desire for domination through the cunning of vision, it is argued, is the archaic archetype of the hypermodern desire to cultivate a pure technological transparency which would make humankind the possessor of the earth by enclosing it. In the hypermodern context, the Apollonian enclosure of city walls becomes a Socratic-technological fortification of communications. This fortification, if it were to be triumphant, would need to eradicate even the threat of unpredictable others, distancing person from person by withdrawing them inexorably into the machine's enclosure.

The appendix, offered as a conversation, refuses to perform a convenient act of enclosure by summarizing and so concluding the preceding text. Instead, it takes the form of an open and personal exchange which obliquely touches on the themes that precede it. What is offered is a compressed, dialogically organized invitation to embark on further thought.

NOTES

1. See Max Horkheimer and Theodor Adorno, *Dialectic of Enlightenment*, ed. Gunzelin Schmid Noerr, trans. Edmund Jephcott (Stanford, CA: Stanford University Press, 2002).
2. See Horkheimer and Adorno, *Dialectic of Enlightenment*, 1ff, 34ff.
3. Jean Baudrillard, 'Whatever Happened to Evil?' in *The Transparency of Evil: Essays on Extreme Phenomena*, trans. James Benedict (London: Verso, 2009), 95.
4. For an insightful discussion of this, see Paul Weller, *The Rushdie Affair* (London: Continuum, 2009).
5. Theodor W. Adorno, 'The Essay as Form', trans. Bob Hullnot-Kentor and Frederic Will, *New German Critique* 32 (Spring–Summer 1984): 161.
6. László Moholy-Nagy, *The New Vision and Abstract of an Artist* (New York: Witternborn Schulz, 1947), 27.

Introduction

1. SEEING, CULTURE, HISTORICITY

Sight requires more than just light and what it illuminates. Seeing always takes place in a world. A world's resources (that is, what it draws upon in order to be a world) do not lie in immediately experienced sensations that imprint themselves directly upon our minds in the form of lively sensory 'impressions'[1] which are turned into life-drained copies of them called '*Thoughts* or *Ideas*'.[2] This view is endorsed by a certain historically powerful model of the philosophy of empiricism, which 'emerged as the paradigm of philosophy'[3] and is associated in the modern era with a specific and dominant reading of the thought of eighteenth-century thinker David Hume.[4] Such a conception of empiricism conceives of it as being foremost a theory about what we know (an 'epistemology', to use an over-worn and over-valued word) which offers an account of knowledge that harmonizes with the de facto interpretative tendencies of the natural sciences. As Gilles Deleuze has powerfully demonstrated, an empiricist like Hume can also be persuasively interpreted in very different terms, namely, as a social theorist who emphasizes not our conceptual abilities but the passionate dispositions of our sensibility.[5] In spite of moments of resistance such as Deleuze's attempt at developing a radical version of it, empiricist thinking, especially in the guise of positivism, frames the modern era as the obedient servant of the scientific-technological spirit which animates it.

One can resist this framing. A world, to return to the point at which we began, is never simply 'there', imprinting itself on the 'empty cabinet' of a mind[6] which passively mimics pre-given sensory forms. To reject this version of empiricism is not to deny empiricism altogether any more than it means rejecting belief in the notion of a concrete environment in which experiences and sensations happen. One obvious example of an analogous approach is furnished by the thought of John Dewey.[7] Environments in all their concreteness are necessary for culture to be possible. Environments, however, signify nothing independently of the cultural imperatives which dictate how we relate to them. We relate to an environment that we see, just like every other aspect of it, in virtue of concrete practices. Practices are ways of doing things (habits, customs, traditions, observances, and the like).[8] The template for such relations is neither empirical-perceptual nor rational-conceptual. It is, rather, practical and engaged. Grasping the nature of this practical engagement facilitates a pathway not only into the nature of seeing and the visual but also into understanding the tendency of archaic conditions to abide with them.

To be in a world, Martin Heidegger comments, entails being locked into it in a manner that, once this condition has been properly understood, has significant implications for understanding visuality: 'If I "only" represent' the things in my environment or even '"do no more" than "think" about them, I am no less alongside the entities outside in the world than when I *originally* grasp them'.[9] The central point here concerns the enabling condition of perspective, a point of view which opens up a world by arranging the environment as a domain of practical concern. One can take as a starting point, as Heidegger does, the fact that we live in a world of equipment.[10] Every specific piece of equipment exists for us only in relation to the fact that it belongs to 'other equipment: ink-stand, pen, ink, paper, blotting pad, table, lamp, furniture, windows, doors, room'.[11] One thus never sees 'things' of this kind as a collection of isolated individuals that, taken together, come to make the 'contents' of a room. There is, to put the matter more bluntly, no such thing as a room-space that is passively waiting there beforehand to be filled up with handy 'stuff'. When I enter a room what greets my eyes is the room 'not as something "between four walls" in a geometrical spatial sense but as equipment for residing. Out of this the "arrangement" emerges, and it is in this that any "individual" item of equipment shows itself'.[12] My ability to get sight of and lay my hands on my pen, for example, relies on the fact that it already exists for me in the context of 'a totality of equipment', i.e. of useful stuff, which forms the environment as one of practical concern.

Equipment reveals itself foremost in our practical engagements with it. Under such circumstances it is never known in a purely abstract 'theoretical' sense. Equipmental stuff cannot, it follows, be exhaustively grasped as an object of detached 'theoretical' contemplation. Rather, whenever putting-to-use happens, what is revealed about a piece of equipment is that there is a kind of knowledge-in-use that is utterly appropriate to it. Heidegger's well-known example of this concerns hammers. 'Hammering' shows itself in the 'handiness' of the hammer. A hammer's handiness is an aspect of it that could not in any way be encountered simply by standing back and looking at the hammer, if by 'looking' one means a detached contemplation of the hammer-thing. 'Looking' in relation to action is never simply looking-on from a detached viewpoint supposedly 'outside' or beyond activity ('theory' as opposed to practice). However intently one might stare at a tool's 'outward appearance', one would never grasp the practical, handy aspect which constitutes the fundamental condition of its amenability to being put to use. Practical activity, it follows, is not some kind of sightless groping about in the world that can be easily contrasted with a purportedly 'insightful', theoretical looking. Practical activity 'is not a blind one; [for] it has its own kind of sight, by which our manipulation is guided and from which it acquires its specific Thingly character'. To see in this context involves a kind of cautionary looking-over in order to get something done, for what is 'ready-to-hand [i.e. amenable to being put to use] is not grasped theoretically at all'.[13]

Equipment vanishes from overt theoretical sight in use and as such is not, when conceived in this context, an object of contemplation—what we concern ourselves with in using equipment is the work at hand. The needle vanishes in the making of the shoe, the saw vanishes in the cutting of the wood, just as the keyboard vanishes in typing in pursuit of the production of words. The human world thus consists of 'things' we put to use. It is, however, also made up of more than merely useful things. The everyday world of doing, Heidegger argues, consists of both equipment which is organized around forms of work and 'those Others for whom the "work" is destined'.[14] Work in this way relates to he or she who will put to use what is made in another's work (clothes, for instance, are made to be worn by someone). Hence, things that are made necessarily presuppose that the world consists of other persons too. The environment itself (which the crude empiricist dubiously conceives of as no more than a domain of brute sensations) is inexorably laid out for us in these terms: 'The boat anchored at the shore is assigned in its being-in-itself to an acquaintance who undertakes voyages with it; but even if it is a "boat which is strange to us", it is still indicative of Others'.[15] The point is that seeing things entails seeing a world not in terms of the material stuff which populates it and onto which Others are somehow grafted, like an afterthought. Others are not superimposed onto the world or inserted into it. The world could not be a world if Others were not there at all for they are an essential component of it as such. 'Others', here, is not a term that denotes something akin to 'other minds' than one's own, i.e. other little consciousnesses from which one is forever barred intimate access. Others 'are rather those from whom, for the most part, one does *not* distinguish oneself—those among whom one is too', for 'the world is always the one I share with Others [. . .] They are encountered from out of the *world*', in which the individuated subject, as a condition of circumspective concern, 'lives'. Others, in short, 'are encountered *environmentally*'[16]; that is, they are always already there in the world as a precondition of its being a world at all. The 'individual' subject thus always encounters itself as being at stake in its dealings with the environment, but the environment is composed of Others, too, such that selfhood never begins with the supposed fact of an isolated self-awareness which then reaches out of itself 'into' the world to try and find others. The philosophical 'problem' of other minds is a bogus one, since it could not arise were it not for the erroneous presupposition that thought resides in an isolated, autonomous, self-conscious 'mind' that is separated from the world it appears to look out on by an unbridgeable gap. The self, in other words, is not essentially an 'inner' consciousness that peeps like a voyeur 'out' of the chamber of its consciousness through keyhole eyes at externally existing things by way of the stimulation of its senses. To see—even to 'see out'—is to be 'there' in the world already.

An approach of this kind, as Vilém Flusser powerfully demonstrates in his *Towards a Philosophy of Photography*, can be developed in the direction of including such things as our imaging equipment, the supposedly passive machinery 'through' which late modernity increasingly records and charts its experience and cultivates its self-image. We tend to think of the camera as a device that passively channels the light that is reflected from the environment into it through the aperture of the lens. But the content of a photographic image is better comprehended as the manifestation of the technical-social scientific order that created the device which 'captures' the image. The modern camera is a device programmed by the 'photographic industry', which in turn can be related back to 'the industrial complex that programmed the photographic industry; that of the socio-economic system that programmed the industrial complex; and so on'.[17] One cannot disentangle the machine which does the looking for us from the conditions which give rise to it any more than one can separate the subject from the world that constitutes the condition of his or her personhood.

The central point here concerns the enabling condition of perspective, a point of view, which opens up a world by arranging the environment as a domain of practical concern. As Charles Taylor points out, such a view rejects the conception of knowledge as invoking the empty universality 'of "the view from nowhere", to use Thomas Nagle's suggestive phrase'.[18] To be in a world, and hence to be 'looking at' as well as 'thinking about' or simply 'touching' it, entails already being *involved with* and absorbed *within* it. One is no less involved when thinks about or 'represents' the world to oneself in a manner that seemingly sets it at a distance (i.e. when one 'objectifies' it in a theory) than when one is acting practically. 'Objectification' and 'objectivity' are standpoints achieved in a world, not above or beyond it. To be in a world means to be immersed in an environment in a condition of concern and fascination, and already to have grasped and interpreted it in terms of this condition. To think about something, to look, hear, or touch is already to stand in a practical relation to the environment that is pondered, gazed upon, listened to, felt, etc. Our concepts, however abstract, are ultimately engagements which betray our having presupposed that we are 'there' in the environed world and that there is something at stake for us in our being there.

The notion of a world, Heidegger also notes, turns on the contrast with worldlessness. 'Things' that are worldless have no world. Consequently, they cannot be independent sources of meaning. Consider a stone. If thought of independently of our possible engagements with it, a 'stone is worldless'.[19] It is a piece of stuff that simply lies there. Light is also worldless stuff. The beams that bathe a landscape do not suffuse it with meaning. Culture does. Nothing that is seen can ever be equated with the worldlessness of the elements that seeing arranges without negating the cultural domain which renders what is seen meaningful. This cultural domain endows visuality with its possibilities of significance. From this it follows that seeing always takes place under the condition of what Martin Jay has aptly called the 'cultural variability of ocular experience'.[20] An observer's eyes are never simply spheres of light-conducting jelly. They scan culturally articulated space.

Media archaeologist Friedrich Kittler's lectures in *Optical Media*[21] indicate how one can chart the emergence of the Western form of this cultural variability in concrete terms. For example, Kittler shows how the Renaissance painter's hand was guided by the conjunction of geometrical rules inherited from the Greeks which formalized principles of linear perspective and refinements in the equally ancient machinery of the *camera obscura.*[22] With the Renaissance perfection of linear perspective, which can be traced back to around 1420, 'all of the lines, corners and proportions of an image appear exactly the same as the image they reproduce on the retina'. This had the decisive consequence of transforming painting into the 'engineering of illusions'[23]—it must be remembered, a 'true likeness' of something or someone in a painting (or, for that matter, a photograph) is an engineered illusion. The *camera obscura*'s use in the Renaissance as an aid to drawing likewise marks a significant stage in the organized automation of optical information, combining 'for the first time the optical transmission of information with the optical storage of information; the former function is already fully automatic, whereas the latter is still manual'.[24]

As Kittler notes, the Renaissance development of linear perspective in combination with the refined mechanical properties of the *camera obscura* modified far older practices—ones which still abide in the contemporary world, e.g. in the standards which govern the manufacture of contemporary photographic equipment.[25] One can, it follows, trace contemporary forms of practice back not merely to Renaissance origins but to archaic roots which reside in the founding of 'a science of optics' in the culture of ancient Greece.[26] In the case of linear perspective, these roots spring from 'the implicit (and later entirely explicit) [ancient Greek] assumption of an infinite universe, which corresponded to an infinitely distant vanishing point in every single perspective painting'.[27] In the case of the *camera obscura*, the ancient understanding of principles first elucidated by Aristotle[28] was further elaborated with the development of Renaissance optics to generate a 'description of perspective [as such] from the perspective of ballistics'.[29] With regard to the latter, one can note the presence here of another all-too-ancient demand, namely, the requirement for effective attack and defence against the threats of enemies in war (a theme that will be returned to in due course).

The (now digital) camera is one prominent modern outcome of this tangle of ancient and Renaissance elements—as in its turn is modern painting, which is likewise replete with photographic elements.[30] When one reaches for a camera, however 'thoughtlessly' or 'spontaneously', to take a snapshot, a range of pre-established visual possibilities and preconceptions to which one already conforms in the very act—that of framing the subject; that of angle of view; that of what constitutes a 'natural' 'representation' or attitude; that of what counts, therefore, as 'worthy'—are already at work in the material structure of the machine and the material relation one necessarily forms with it. Seeing, in this constitutive sense, has always already happened before any specific 'moment' recorded as an image: the demands of the machine and the inherited domains of practices which accompany its being put to use are implicit in both the making of the image and the gaze of the subject which seemingly guides such making. In this specific sense, it is hard to agree entirely with Walter Benjamin's claim that 'mechanical reproduction' delivers us from a 'parasitical dependence on ritual' such that works of art rather than 'being based on ritual' come 'to be based on another practice—politics'.[31] This is not to deny the political possibilities that the image-making machine initiates nor to question the degree to which the camera operates 'under a new law' which, as in the case of cinema, slices up the environment according to dictates which serve to reveal an 'unconscious optics' at work.[32] It is simply that what the device unleashes by way of a politics, a new law or the unconscious is never wholly 'new', for even the machine is a bearer and conveyor of a history which involves modes of evaluation that precede it and coalesce not so much 'within' it but *as it*.

One can take this relation between device and what Kittler identifies as the device's ancient but living past as a starting point for critical reflection.

Speaking like Günther Anders, one might say that in such moments we discover ourselves precisely in virtue of our feeling in thrall to a machinery which, even as we seek to see 'with' it, seems to see us, so generating a disturbing reciprocity between tool and person, between the 'it' of the non-homogeneous living self and the 'it' of the tendency of machines to absorb other bodies into their rhythms.[33] One might in turn venture the proposition that 'simply' seizing hold of even the most 'contemporary' of imaging devices thus entails grasping along with it the historical-cultural world concatenated into relations of bodies and machines which simultaneously produces and inheres within it. History, one might say, is already there prior to our picking up the camera as a condition of doing so, fashioning the visioning of the present and, consequently, also the future in the material conjoining of practices, bodies, and devices.

To see is to construct. To construct is to follow inherited imperatives of observance (conventions are never wholly one's own) which allow one to observe (to attend to something). In this sense, any world is essentially historical.[34] A world's historicality accompanies it as a condition of its possibility in the present. Humanity's 'own past', in consequence, 'is not something which *follows along after* [. . .] but something which already goes ahead of it'.[35] The same can be said in relation to technology for it is entwined with our historicity. It, too, displays the tendency to lie ahead of us as a condition of envisioning the future at all. What 'future' can we think of that is not also replete with its fabulously unknowable yet dimly imaginable mechanical assemblages?

Any world is, for this reason, always Janus-faced. We are obliged to think in terms of a temporality which returns us to the past continually even as we look 'ahead'. The rippling skin of a distant body of water may intimate something lurking beneath, but the significance of that 'something' depends upon the historicity of one's cultural world, that is, upon the engagements, memories, fears, and hopes possible within it. It is impossible to disentangle these elements from one another, apart from as a thought-exercise in abstraction which comes at the price of denuding the world of its concreteness. Tracings of clouds in a sky can point like weather-cocks to ill winds, and calm waters can be auguries of an easy voyage home, or either can stand as invitations for scientific or aesthetic reflection. One does not 'see' rippling waters or traceries of cloud with timeless, 'naked' eyeballs any more than one ever simply 'hears' with only the vibrations within ears or 'feels' with the fleshy strings of one's nerve fibres alone. The unfolding cultural drama of sign and language in which the visual and its possibilities are formed has been laid out for the individual in advance: it is always there beforehand and alongside us. This is not a merely 'individual' or 'subjective' condition of awareness but a shared, pragmatic, cultural condition because the shared past is what endows subjective awareness with its possibilities.

The 'present', understood in these terms, is a temporal amalgam. Inherent in the inception of any present time (any 'now') is a necessary tracery of archaic cultural elements of practices and practical engagements that are latent within it. We identify and make sense of the environment in terms of these 'traces', to use a word strongly associated with Jacques Derrida, as much as we do in relation to our 'current' concerns. As Derrida notes, this tracery of elements can be understood as an 'arche-phenomenon of "memory," which must be thought before the opposition of nature and culture, animality and humanity', and the like, is possible.[36] The arche-phenomenon, in other words, is the archaic yet constantly present force that opens out the pragmatic possibilities of sense which make up any historical world. Out of this shared pragmatic condition of human historicity spring the inheritances of words like 'I', 'you', 'we', 'they', and 'it'. By the same token, terms like 'see'[37] always resonate with a historical dimension whose presence relates contemporary, shared ways of doing things to the inheritance of cultural memory (and do so even in the case of repression of memory).

The 'seeing' and 'vision' characteristic of any field of vision and its cultural domain is thus an inherited 'historical construction'[38] no less than any other aspect of human experience. This condition permeates the realm of the visible not least because it inheres in practices of seeing. To 'look on' a spectacle, for example, is to participate in a sphere of drama and ritual which conjoins perspectives ancient and modern, be it that of the spectator of the classical drama and the viewer of the contemporary television soap opera or the cheering onlooker of the gladiator or bull-fighter and the watcher of the mass-produced, televised contest voicing opinions via the latest smartphone. Sometimes one sphere of practices can bleed into another—the person before an artwork or abomination, camera-phone in hand, can mimic a worshipper before a shrine . . .[39]

2. SEEING, PRACTICES, PURPOSES

Seeing has just been linked to the pragmatics of ways of doing things; to the cultural worlds that are constituted by these; to spheres of gesture, sign, and language; and to archaic historical inheritances. Practices are bound up with our understanding of the past, with history. This assertion of the power of the archaic does not, however, licence the idea that what is older (i.e. most ancient) should be taken to be more 'authentic' because it is somehow 'closer' to nature, to the 'origin' or 'order of things'. This trajectory of thought, pursued influentially by Leo Strauss, would have us take the archaic as signifying an autonomous and unchanging source of wisdom in the form of the notion of an original, natural law revealed through a philosophy of first principles.[40] There are no 'first principles'. One can elaborate this last claim by turning to some comments Nietzsche makes in *On the Genealogy of Morality*. These comments suggest a way of grasping how the archaic erupts into contemporary understanding as a contingent but powerfully wrought constitutive element. This eruption, in turn, is related to seeing and the domain of the visual, as will become especially clear in the following consideration of Nietzsche's conception of power as involving the marking and scarring of bodies, institutions, and practices.

It is a common mistake, Nietzsche argues in the second essay of *On the Genealogy of Morality*,[41] to assume that beliefs, concepts, things, customs, or organs 'develop' in the sense that the word is usually taken to suggest. It is tempting to think of 'development' as implying a purposive kind of evolution in which a thing changes from one state to another guided by fixed guiding principles that explicitly link every stage in its development to what precedes that stage and realize themselves concretely in the process. To think thus is to be gripped by what John Dewey calls the 'intellectual fabrics' of pre-Darwinian thought.[42] However, a thing, custom, or organ, Nietzsche argues, is best regarded as an artificial whole, a synthesis of disparate elements lacking rational connection (their relations are 'synthetic', i.e. they cannot be deduced from one another by invoking logical rules) whose unity as an ordering of parts is not susceptible to being explained by the attribution of an overarching purpose.

Nietzsche furnishes a useful and provocative exemplification of this claim by way of reference to the eye. It might be tempting to think that the eye developed in order to facilitate seeing since seeing is the function we primarily associate with eyes. This, however, would be to presuppose a world in which a pre-existing eyeball-idea guided the eyeball's emergence into being. A conception of this kind entails being committed to a disembodied notion of visuality: a seeing-without-organs body. But to think this one would have to hold that seeing came before eyes did, which is absurd.

The same point can be extrapolated so that it covers such things as human customs, rites, traditions, procedures, and conventions—all of which presuppose a realm of visual observation as at least one of their essential aspects. Customs, rites, traditions, procedures, conventions, and so forth are modes of observance insofar as they presuppose the participatory gaze of various witnesses (including those who 'observe' them). They are, in this regard, inexorably visual events which guide the eye and, in guiding it, draw the observer into themselves. The eye moves guided by conventions. Only then, in the act of conforming, does it 'see'. Who am 'I' under such circumstances? Answer: One who observes and in observing has already conformed as one of observation's preconditions. 'Observing', in this regard, displays the double implication of 'looking on' and 'according with'. Customary elements of this kind, which seem merely to be contingent features of meaningful acts, are in fact always older (usually much older) than any guiding purpose that happens to be associated with them at one time or another. This is the case whether it might be the rituals of a religious service; the opening of a parliament; the wearing of faintly ridiculous wigs by British barristers in courtrooms; the facing up of opponents in boxing rings, football fields, or battlegrounds; or the simply casual manners of our greeting one another. Customary elements and conventions are no mere 'accompaniments' which simply frame acts. They are constitutive conditions of the possibility of doing things which tend to outlive the ends they supposedly merely frame. The supposed 'frame' in this way guides eye and hand and mouth as an essential condition of directed meaningfulness. Moreover, procedures (like bodily organs) endure in relatively unchanging forms where the different and various purposive meanings associated with them come and pass away. Procedures pattern behaviour and haunt the possibilities of sense which can be articulated by way of them. They encircle actions and in doing so permeate them.

The above entails abandoning any commitment to the view that history is guided by overarching teleological principles. The notion of teleology rests on two related beliefs. The *first* is the presupposition that a thing's development can be taken as exhibiting the feature of being guided toward some end, i.e. that some kind of purpose is at work within the thing or in relation to it that constantly shapes it and which is evident in that shaping. The *second* is that this purpose is 'rational' in the sense that it is the 'logical' realization of a goal and that it is therefore always governed by the imperative which demands it take the shortest, most direct, and most efficient route to that goal's fulfilment. The latter conception of acting supposes that 'being rational' means avoiding doing things in ways that do not lead to anything 'beneficial' with regard to predetermined outcomes. The 'rational' person does not squander his or her time or money, does not idly cast his or her eye over things independently of some stipulated end ('target'), is purpose-driven ('strategic'), etc. This received teleological view of development thus presupposes a kind of economy to be at work in nature and human action alike which maximizes output by minimizing the expenditure required relative to it since such expenditure is conceived as being a means to an end and any means which does not focus on an end is waste.

History, however, does not abide by this rationality any more than nature seems to.[43] This is suggested by the fact that, for one thing, something can always arise from what seems to be its opposite. The origins of reason, for instance, are tainted by unreason, logic by illogic, morality by 'immorality', truth by error,[44] and the eye originates in sightlessness. The history of a thing is not characterized by linear development, by a fixed perspective which guides it toward an invisible but sure goal in the future, just as the parallel lines of linear perspective supposedly point with surety toward the infinite place where they must finally meet and unify. Rather, a custom, an organ, or suchlike can be read as the manifestation of a series of realigning moments enacted with varying degrees of violence in which the various relations of the parts that constitute it are altered, sometimes radically.

This has some important implications. First, the history of something when regarded in this way becomes 'a succession of more or less profound, more or less mutually independent processes of subjugation exacted on the thing'.[45] Historical change, in other words, is exhibited by way of the distorted and sometimes concealed traces of transformations left when something has been subjected to a commanding power that has overcome and remoulded it. A commanding power leaves these traces as a series of often near-invisible marks, scars, and patterns across the body of a thing, a tradition, a habit, or a procedure. Second, the relation between the characteristics of a thing and what seeks to subjugate it is not passive. Anything that is set upon responds to imposition with resistance. To take a supposedly brute example, one can think of the practice of metalwork. Iron in being shaped resists the hammer. Its traces of resistance are evident in what it is amenable to being fashioned into and what not. The very pliability of iron, in this regard, is a trace of its resistance made visible yet concealed in the finished form, which seems to pay testimony only to the power that moulds it. Sometimes, of course, resistance is largely futile and the thing is subjugated or incorporated into a wider cultural domain which in turn moulds itself in relation to that pliability (the 'Iron Age'). Sometimes it is not. Either way, some things can be altered as much by reaction to an attack as by the attack itself—even a failed attempt to overcome something can leave in its wake decisive traces and alterations, like the scars that might pattern a resistant door or gate from the blows that have tried to break it down.

Whatever outcome might turn out to be the case in any particular instance, implicit within Nietzsche's view is the contention that there are no a priori rules of history. Contra someone like Leo Strauss, there is no 'nature' composed of moral rules for history and the agents which populate it to deviate from since no nature is given in the first place in law-like form beforehand. Things are shaped by encounters devoid of any relation to a universal conceptual ordering that might be assumed to be at work within them. What drives historical 'development' is subjugation, either as threat or attempt at subjugation, or as the countering of it by resistance. Historical fields are thus multiple terrains consisting of syntheses and counter-syntheses of contending elements. The essential point here is that history is not exemplified by bodily interactions which conform to a pre-ordained conceptual ordering according to objective 'laws' of any kind but by *activities* which give rise to conceptual and other forms of cultural order. Another way of putting this would be to say that history is not driven by eternal, autonomous, purposive 'ideas' which reveal themselves in its events and spectacles. An ordering of concepts and ideas is in any given case a product of the active interplay of contending forces relevant to it. This play obeys no rules given independently of particular encounters. What we like to think of as historical and other 'laws' arise from these encounters as signs of their having somehow come about, as symptoms of one form of hierarchy triumphing, in the form of cultural achievements of understanding (since such achievements like to put labels on things and events in order to 'explain' them).

We are presented by way of Nietzsche's account with two related aspects of historical existence. There is a thing's 'form' and there is the 'sense' or 'meaning' (namely, the guiding feeling of purpose) that happens at any one time to be associated with that form. Both form and sense are fluid and so susceptible to continuous revision and re-formation. The 'meaning' which endows a sense of overarching purpose, however, is far more fluid than the form. The form, in other words, is far more enduring than any purpose-directed meaning could be. This is one reason why the cultivation of traditions is such an essential feature of cultural life: traditions, customs, and the like ensure the endurance of the possibility of meaning, of a sense of direction to life, albeit not this or that particular meaning. That meanings are sought which supply purposiveness is a necessary feature of cultural life; but which purposes one ends up with at any one point is a contingent matter. Another way of putting this would be to say that like bodily organs, customs and the things related to them have a tendency to alter less rapidly than the purposes people usually feel are so obviously and inexorably tied to them.

Our tendency to prioritize meanings which endow a sense of overarching direction is also a tendency to see the world upside-down. We think meanings of this kind matter most, whereas they are ephemeral in comparison with the supposed 'vehicles' which seem merely to convey them. It could be that this is a precondition of our 'seeing' interpretatively at all, no less than the mind's inversion of the image projected onto the retina is a precondition of the world looking the 'right way up'. Practices may seem coincidental. However, they endure with relative (not to say dogged) constancy due to their slowness of transformation, while apparently stable meanings slip by at speed, like apparitions. Practices are, in this regard, amenable to being thought of as cultural 'organs'. The human heart is an organ whose form might well change only imperceptibly over the life of the species, but the meanings associated with it shift wildly: the heart as the site of the soul; as closed-off, private, 'inner' space resistant to the intrusive gaze of others; as heat source; in the modern era, as a pump for circulating the blood; etc. Traditions and things, too, tend to endure in like manner more doggedly than the purposive meanings that accrue to them.

The desire for meaning that endows a general sense to states of things may endure, but such meaning is never seen independently of the procedural doing. Any particular meanings that endow a sense of purpose, Nietzsche holds, can only rule for a time. They must come and go.[46] There is, above all, no general principle which legitimizes the belief that the association of a thing or tradition with some purpose represents anything more than a contingent and passing historical moment. The usually accepted functional 'meaning' of any 'thing' and the origins of that thing are best regarded as being different in kind. That a thing, custom, or organ pertains at one moment in time to a certain sense reveals that a dominant interpretation (a dominant form of power, a dominant modality of purpose, a commanding ordering of social relations which exemplifies a 'form of life', to use Wittgenstein's phrase) which is, moreover, often concealed by it, is at work and putting it to use. The contingent sense tells you nothing about the origin of the thing, custom, or organ which almost certainly came about for a constellation of unrelated reasons not all (and possibly none) of which are connected to one another in a way that can be grasped by the imposition of a currently dominant, unifying sense.

Another way of putting this is to say that the notion of a universal perspective on things is always lacking in spite of the appearance to the contrary. The notion of universality in this sense is, after all, dependent upon the idea that there is a standpoint from which one can see, survey, and so enclose everything within the ambit of a single look. It is procedures, however, which persist—often in the form of variations. Their persistence brings with it echoes and afterimages of earlier but now largely alien webs of relations between things, earlier purposes, and relations of domination, which inhere in them in ghostly form. Such ghostly aspects often exude an aura of familiar unfamiliarity that it is easy to ignore—like a commonplace word whose etymology one does not for a moment pause to ponder or the furniture in a long-familiar room whose taken-for-granted arrangement stretches back beyond one's earliest childhood memories to the conventions and habits of others that render such individual memories possible.

In *On the Genealogy of Morality*, Nietzsche gives examples of the shifting historical valances of different senses associated with more-or-less-stable procedures and practices. For most readers of the text, the obvious instance of this is furnished by his account of the origins of the evaluative habits associated with moral talk (noble and slave morality) which is suggested by the etymologies of words.[47] Moral talk, for Nietzsche, like any other realm of practices, is synthetic: it is hybridized through and through as a historical bundle of often incompatible meanings and purposes signifying struggles. This is why it is a mistake to view morality as a unity which pertains to a meaning independent of the domains of practice associated with it. Viewed historically, 'morality' is a tangle of moral meanings, of differing and contending ends, bound up with relatively enduring ethical procedures which, because they are enduring, become the domains of social struggle in and by way of which the purposes are asserted. The meanings that coalesce in this way cannot be disentangled from one another or stripped away in order to reveal a timeless, residual kernel at their core. It is impossible to show what 'good' 'really' means (that is, what is univocally 'good' 'in itself') since the historical unfurling of the practice inextricably interweaves competing and sometimes incompatible evaluative elements. The original 'meaning' of evaluative terms is significant when it comes to understanding moral norms because confronting it undoes the presupposition that the same meaning and purpose of the 'good' is at work in all moral discourse—or the lazy attitude that this or that current, generally accepted notion of 'good' is the right one. The falling away of origins from current dominant meanings reveals the essential contingency at work in the ongoing development of the practice since meanings shift in harmony with the inharmonious and fluid historical conflicts of power relations.

However, the most significant aspect of historical practice that Nietzsche considers in *On the Genealogy of Morality* is not the kinds of moral talk associated with noble and slave. It is something bound far more closely to the shaping of the visual. It is the webs of practices associated with exchange, creditor, and debtor. It is out of these, he contends, that the contested field of ancient moral-cultural practice that modernity inherits subsequently unfolds.[48] Exchange begins with the primitive relation between a biological individual (a proto-human devoid of the self-understanding one takes for granted when considering a 'person') compelled by dangerous circumstances to take refuge within a group (i.e. a community). Driven together in this way, each individual becomes bound and regulated by the demands of communal life. These demands, in the case of the 'human animal', get articulated as social practices. For example, desires and drives which damage the community come unthinkingly to be curtailed by powerful, ritualized, punitive measures.[49]

Nietzsche's claim is that once this communal structure is in place, much else follows. For one thing, there arises in tandem with communal terrain an exchange structure organized around the observation and memorizing of observances linked to duties and notions of equivalence, compensation, credit, and debit which determine the extent of a punishment by stipulating what is to count as repayment for a wrong: 'An eye for an eye'.[50] Here in measuring practices lie the founding conditions of culture,[51] including the webs of ways of doing things through which persons come to self-understanding.[52] Here also reside the conditions from out of which organized social (rather than merely communal) forms of existence get articulated, e.g. in the sequence of initially custom, then law, then the state, etc.[53]

Nietzsche's argument is that procedures (habits, customs, traditions, practices) acquire a compelling nature by way of their association with communally cultivated feelings. Ways of doing things (including thoughts, for thinking is no less a practice than other forms of observance[54]) are orientated around conditions which are essentially shared and in being shared are also embodied, visible, and tactile rather than mentally abstract and 'conceptual'.[55] It is this entwinement of the visible, the tactile, and the habitual into a synthetic structure (what one might call a technology of practices) which makes each of them compelling and which comes to endow them with their 'oughtness'. The realms of culture which open out with the cultivation of procedures depend on this concrete, bodily engagement and coordination. Culture takes the form of shared norms: norms depend upon the propensity wherein habitual observances articulated in movement and gesture initially form and then continue to shape and re-shape the shared environments where communities spring up and from which selves emerge endowed with self-understanding.[56]

Human visuality is conjoined explicitly by Nietzsche with this primitive and founding cultural articulation of the habitual tendency to comparison, measuring, and calculation. The extent to which humans evaluated their environment by way of unreflective adherence to such propensities had a decisive bearing on how they saw (that is, looked upon and so understood) both it and themselves.[57] In other words, sight was cultivated as a disciplinary inclination to follow certain procedures linked to exchanging and measuring: 'The eye was now set for this perspective', and, driven by 'the ponderous consistency characteristic of the ancients' way of thinking [. . .] never deviated once it was moving' so that rapidly enough came 'the great generalization: "Everything has its price, everything can be compensated for"'.[58]

It is not difficult to notice an important implication of Nietzsche's contention here. The eye never simply 'sees' in the sense that one can conceive of it as a purely receptive organ lying passively in wait for illuminating rays of light. In the case of creatures like us, the eye is receptive precisely to the extent that it is already trained to be active, constructive, on the look-out for meanings by way of being orientated to the environment by practices linked to estimating by way of measurement. Our gaze, in short, is directed by the unconscious inheritance of practical, procedural contexts. As such, it is always already evaluative. This condition of being 'always already' orientated toward the environment is what makes goals (purposes, meanings) possible. To calculate, for example, means to approach the world with a view to its amenability to being counted and divided up accordingly: to see the world as a space which is patterned by number and thereby rendered predictable. In turn, such a mode of seeing allows one to pursue ends . . . to count in order to make sure one's sheep have not wandered off or been stolen, 'Yan, tan, tethera . . . '[59]; or as a means of symbolizing the divine; or, today, to calculate how many cells in a sample are cancerous—today, also, to count so as to accumulate data in order to see patterns of 'individual' behaviour.

The gaze is set by cultivated bodily propensities which unfurl within networks of historically inherited observances guided by the tendency to contrast and compare, count and reckon, equate and differentiate. As John Berger's work suggests, such propensities are powerfully at work with regard to the manner in which images construct identities. The supposedly 'innocent' visual pleasures of high art conceal, amongst other things, sexual domination.[60] Images of women betray the ancient power that men exert. Representations of male and female populate and so structure the visual in telling terms: a 'man's presence is dependent on the promise of power which he embodies', whereas 'a woman's presence expresses her own attitude to herself, and defines what can and cannot be done to her'.[61] One is the bearer of the gaze, the other its passive object. Thought of in this way, 'the male gaze', to use Laura Mulvey's influential phrase,[62] emerges as the opposite of something 'objective' or 'disinterested'; it is subordinating, radiating power. To approach the visual in this way is to understand how patriarchy and other ancient forms are preserved and replicated in contemporary life by the inheritance of practices of looking which find expression in publicly distributed images of all kinds. 'Today', Berger concludes, having surveyed the tradition of European painting, the patriarchal 'attitudes and values' which 'informed tradition are expressed through other more widely diffused media—advertising, journalism, television',[63] and, of course, more recent media. Movements of the eye are already focused and trained to scan the environment in accordance with a legacy of cultural inheritance which finds itself repeated and reinforced in the visually articulated domain of contemporary daily life. What takes on the everyday appearance of being something harmlessly pleasurable, namely, 'just' looking for 'entertainment' ('scopophilia'—enjoyment in looking, which Mulvey asserts (somewhat problematically) is a 'primordial wish'[64]), conceals an invisible domain of practices that form and fashion the self and whose power resides in their very invisibility in deep, unthinking habituation and familiarity.

A gaze engenders subjectification and so colonizes those toward whom it is directed.[65] The dynamic of this phenomenon is well captured by Michel Foucault's conception of 'panopticism'.[66] The panopticon encapsulates a power that is quintessentially visual. It is a structure designed so that those contained within it (be they prisoners, psychiatric patients, workers, or children) are permanently subject to the gaze of an invisible observer. This invisible but pervasive look 'insidiously objectifies those on whom it is applied', with the effect of generating 'in the inmate a state of conscious and permanent visibility that assures the automatic functioning of power'.[67]

Resistance to this gaze can be found in disruptively turning it against itself.[68] Following Hélène Cixous, one can begin by challenging the imagery at work in oppositions by occupying their space of operation on different terms—speaking like Nietzsche, by introducing new, contested purposes into the historically articulated space of practices that itself made them possible: the subjected self, once subjected, can speak of subjection and make of it a weapon of resistance. Thus, for Cixous, the opposition 'Man'/'Woman', which when 'we read or speak' unfurls 'the same thread or double braid' mapping out a pathway 'throughout literature, philosophy, criticism, centuries of representation and reflection',[69] needs to be exposed in its complicity with the 'premise of women's abasement'.[70] In the modern era, 'an age where the conceptual foundation of an ancient culture is in the process of being undermined',[71] one does not seek simply to abandon that pseudo-foundation in a vain attempt to reset the calendar to zero by initiating the search for another 'truer' one (this would, after all, be a quintessentially 'male' gesture, colonizing in every sense). One pursues, instead, the path of transformative retellings of the history of the practices in which the notion of the 'foundation' itself (i.e. the ideological belief that there is a purpose sanctioned by 'nature' independently of our practices) is made manifest and thereby placed in doubt.

'All concepts', Nietzsche argues, 'in which an entire process is semiotically concentrated defy definition; only something which has no history can be defined'.[72] History is marked by the generation of a multiplicity of conflicting perspectives in the context of practices which come to be put to use in differing hierarchies of social relations—including the resisting and questioning of them. Insofar as we are all the observers and (usually unwitting) inheritors of various historical elaborations of practices, we all embody many contradictory standpoints. Put another way, our habits of thought are never entirely 'rational', which is why 'not only the reason of millennia, but their madness too, breaks out in us'.[73] It is for this reason that, if one wants to consider the present with a critical eye, one must not look, as Leo Strauss does, to retrieving a timeless moral meaning with which to judge it. A fixed and 'natural' meaning of this kind, which purports to endow everything with a direction from the outset, is a chimera. One can, rather, follow the suggestion made by Heidegger's thought and engage more closely with the legacy of modernity's historicity and its dominant hierarchy of sense by exposing the lingering, rearranged, and reinterpreted archaic forms of practice which both constitute its origins and continue to haunt and shape it. To ask about origins in this way is to question dominant meanings by opening up disruptive inheritances at work in contemporary practices in order to expose archaic aspects still at work within them and so question the authority of the present from within.

Over the course of time purposes get radically reinterpreted by being subjected to different demands (i.e. different relations of power). Likewise, practices of vision (how the eye is directed by understanding shaped by procedures) bear the historical traces of these sometimes sudden and violent reinterpretations. As a child of understanding, the eye's is a scarred historical heritage. As we look on we cannot help at the same time, like Janus, looking back. One cannot 'see' independently of evaluative feelings, that is, without being subject to the heritage of procedures which give rise to such feelings and whose traces inhere in modern no less than so-called primitive modes of life. This, one might say, is modernity's great insight into itself. Unsurprisingly, therefore, Nietzsche, himself a true modern,[74] holds that in modernity an archaic inheritance of traceries of enduring sensibilities abides. Primordial feelings rooted in deeply ingrained, evaluative habits compel us to 'draw the conclusions of judgments we consider false, of teachings in which we no longer believe—our feelings make us do it'.[75] Our feelings are bound to our practices, to the old dance of body, spectacle, and gesture which in being gazed upon gives thought its concreteness. The more, it seems, we are self-consciously modern, the more we become aware that we are also bundles of ancient, habitual feelings and observances dressed up in industrial-technological garb. Everything and everyone modern is also always already primitive. This epitomizes modernity's divided insight about itself, in relation to which 'Darwinism' is perhaps only the most obvious and contentious of symptoms.[76]

3. SEEING, MODERNITY, THE ARCHAIC

As has just been argued, we see through a tangle of archaic elements. We moderns never see as 'pure' moderns. Modernity is the peculiar configuration and awareness of this tangle. Modernity is foremost the era of the burgeoning of technology—but it is in this most contemporary of guises that its ancient elements come starkly to light. Technology's fulcrum is the sphere of our perceptive abilities, the domains of the organs of sense, and not least the relatively stable form of the human eye and the practices of seeing which have accrued around it. The means (the 'medium' of technology) invades and constitutes the end. As Kittler says, 'Media determine our situation'.[77] Or, in the words of Mark Hansen, 'Technologies *underlie* and *inform* our basic "ways of seeing" the world'.[78] Power is seen at work when one ponders modern articulations of practices of seeing, not least because a significant feature of modernity's beginnings and subsequent development is the specific technologization of vision which characterizes it.

Speaking in terms made famous by Marshall McLuhan, one can say that with modernity it becomes impossible to ignore the extent to which 'the medium shapes and controls the scale and form of human association and action'.[79] The electric light, to cite one of his examples, appears to the untrained eye as a 'pure medium' devoid of meaningful content. We use artificial light for a multiplicity of purposes and are, in the midst of use, unaware of the way in which it interacts with and moulds the visual. The particular spaces we choose to turn it toward and illuminate seem to be a purely coincidental matter. But the 'conventional response to media, namely, that it is how they are used that counts, is the numb stance of the technological idiot'.[80] This is because every medium initiates transformations of social space (with changes 'of scale or pace or pattern'[81]) which have the consequence of generating new social forms and possibilities. One only grasps artificial light's function as 'pure information' when one appreciates that this function itself is what changes the world: 'The message of electric light is like the message of electric power in industry, totally radical, pervasive and decentralized. For electric light and power are separate from their uses, yet they eliminate time and space factors in human association exactly as do radio, telephone, telegraph, and TV'.[82] Our senses are always organized around and through media. This, as McLuhan notes, is not a recent phenomenon. The European model of the city, for instance, emerged in ancient times in tandem with the development of the medium of phonetic writing. The latter, in differentiating between sight and sound, made possible the elaboration of ancient imperial territory (Rome) into a compressed and easily regulated domain of visualizable order.[83]

As Jonathan Crary argues, modern media, in the shape of technologies of perception, came to the fore in the nineteenth century. Their arrival is borne witness to by the way in which they reconfigured practices associated with sight, transforming not only the sphere of seeing but also the subject who put such 'seeing' 'to use'. Industrialization lies at the heart of this transformation. Nineteenth-century optical equipment took on a new relation to the subject. This relation to the subject, Crary points out, is present within Marx's description of what occurs when a tool is incorporated into a system of mechanized production. For Marx, with the arrival of the industrialized factory, the worker ceases to be the traditional artisan for whom tools are subordinated to the requirements of individual skill. Instead, he or she is increasingly absorbed into a systematically integrated world in which equipment becomes harmonized with the mechanistic demands of mass production. This absorption marks a point where the relation between worker and tool is radically altered. Where the artisan's use of the tool complements the particular demands of work, the factory worker, in contrast, is obliged to conform to the demands of the machine. A worker in thrall to a machine is reconfigured: his or her subjectivity refashioned through integration into a system of mechanized production replete with its dominant forms of organization, such as the structured working day which harmonized the worker's relation to time through the regulating power of the factory clock.[84]

The same goes for machines of entertainment that titillate the gaze, such as the stereoscopic optical devices which became popular in the 1830s. These machines reconstituted earlier modalities of vision by subordinating the viewer to mechanical demands that reconfigured the visualization of the environment. With the stereoscope, for instance, each eye looks separately through a single eyepiece into a manufactured space that is a 'technical reconstitution of an already reproduced world fragmented into *two* nonidentical models, models that precede any experience of their subsequent perception as unified or tangible. It is a radical repositioning of the observer's relation to visual representation'.[85] The point is that technology does not just change what someone can see, e.g. at the level of detail such as is offered by the increased magnification of the microscope. More importantly, it alters a person's understanding of the environment, reconfiguring everyday space and thereby altering the individual's sensibility with regard to their awareness of their own place within that space. This, as Siegfried Kracauer notes, is a key feature of the impact of

> modern photography [which] has not only considerably enlarged our vision but, in doing so, adjusted it to man's situation in a technological age. A conspicuous feature of this situation is that the viewpoints and perspectives that framed our images of nature for long stretches of the past have become relative [. . .] The same applies to phenomena on the ideological plane [. . . so that . . .] photography has effectively impressed upon us the dissolution of traditional perspectives. Think of the many prints picturing unwonted aspects of reality—spatial depth and flatness are strangely intertwined, and objects apparently well-known turn into inscrutable patterns.[86]

The day-to-day world, as Kracauer points out, 'is not born out of the immediate observation of the given', for the concepts and beliefs we cleave to on a daily basis 'are not experiences; they are common currency'.[87] Technology renders questionable the value of this currency, reconfiguring the everyday, changing our relation to what we do by modifying how we do what we do in subtle ways that are both continuous and yet also in some manner discontinuous with what went before. Yet this modification cannot break entirely with what precedes it. It is this ambivalent relation—which Walter Benjamin articulates in terms of an undermining of the power of tradition[88]—which constitutes one of the decisive conditions of the cultural production of meanings in modernity.

Ways of doing things endow a sense of self. They are a necessary (albeit not sufficient) condition of subjectivity. By acceding to a practice I reveal to myself just who I am in virtue of doing so. If I am a good neighbour, for example, I subscribe to the practice of neighbourliness (I will not bully my neighbours or treat them as less than human), and it is only by acceding to the authority of the practice that I can form a valid conception of my own status as a 'good' or 'bad' neighbour. Today, technology affects even neighbourliness since it reconstitutes the domain of possible actions available to me with regard to my neighbour—it may, in the guise of communications technologies, even invite a revaluation of just what constitutes a neighbour or being neighbourly since notions of proximity, distance, and the personal are transformed by the visual and other possibilities afforded by communications devices. Since ways of doing things are constitutive of subjectivity and understanding and modern technology figures itself most essentially into our ways of doing things, it engenders reconfigurations of understanding and our sense of selfhood. Our practices abide, but in their inhering they are transformed. As our practical everyday world changes, so do we and, in turn, this changes how we see it and ourselves. Given this, the person who sees in virtue of being endowed with an understanding increasingly mediated by technological devices which are subject to constant revision and refinement by way of the production process is likely to become as fluid and changeable as the developing technical means through which seeing and the seen become increasingly constituted.

The nineteenth century initiates the pervasiveness of this distinctively modern, fluid condition of self-awareness. The sense of fluidity is well captured by Thomas De Quincey in his 1849 essay 'The English Mail-Coach'. De Quincey contemplates the vanished era of the horse-drawn mail. 'The modern modes of travelling' by mechanical means, he comments, do not compare in terms of 'grandeur and power' with the wind-blown experience of the mail-coach ride when one sits on an outside seat, for although the former possess 'more velocity', their speed is only 'a fact of our lifeless knowledge [. . .] because somebody *says* that we have gone fifty miles in the hour'.[89]

Yet however abstract one's grasp of the machine's speed may be, it cannot be denied that its 'velocity' ultimately dissolves the stability of everything else in its wake: 'when I look back upon those days, it seems to me that all things change—all things perish' as the once thrilling experience of close-up speed on the old coach is now washed away amidst 'the fluctuating mobs of a railway station'.[90]

The notion that the apparently stable things of daily life have become inexorably temporary is, of course, borne witness to most notably by Charles Baudelaire whose first 'novel, terminologically peculiar use'[91] of the term 'modernity' was formulated in 1863, not long after De Quincey expressed his lament for the lost glories of the windy coach ride. Modernity, Baudelaire comments, is 'the transient, the fleeting, the contingent'.[92] Georg Simmel, writing six decades after De Quincey and nearly four after Baudelaire, likewise echoes this when he refers to the 'inner meaning of the specifically modern life' as turning on the metropolis's 'swift and uninterrupted change of outer and inner stimuli'. This constant movement is dominated by a 'crowding', rushing upsurge of 'images' that generates 'the sharp discontinuity in the grasp of a single glance'.[93] Modern city space, in other words, is dominated by unstable conditions which reveal themselves most forcibly in vision. Vision is cut up; as the culture of the city environment offers one heterogeneous stimulus to the eye after another, habits of looking are overwhelmed.

What comes to the fore in Simmel's description is the ubiquity of mobility and its social consequences, amongst which is the intensified role of images in public spaces (not least in advertising). George Grosz offers an interesting example in relation to the latter. In a short piece written during World War I, the powerful, optical nature of modern advertising is linked to the need to impress the seller's message upon the train passenger who speeds by for a brief moment only. In the modern environment, it is one thing to possess the characteristics of the traditionally 'cultured' person, i.e. to know the works of Schiller and Goethe. 'Culture' in this sense is signified by one kind of speed—it is something slowly acquired, a matter of an altogether older tempo of existence which guides the eye. The pressing question now, however, is 'But can *you* ride a bike?'[94] Grosz's question captures well the way the fluidity of modern metropolitan life dissolves the pre- and early-modern illusion of a stable and fixed point of view on the world. To cope with the contemporary environment, the gazer, like the person who negotiates roads and pathways with the malleable mobility of a bicycle or who hurtles past things on a train, must become as mobile-eyed as it is. The viewpoint of the viewer, like that of the once fixed and stable but now portable *camera obscura*,[95] has become transient and the viewer experiences him- or herself as such. The traditional 'split between the perceiver and the world'[96] in which the perceiver is conceived of as an entity looking outward from a fixed, monocular perspective 'into' the 'external' environment is sundered in response to the demands of 'modernization' brought about by 'a deterritorialization and a revaluation of vision'.[97]

Modernity's distinctive feature is the disruption of traditional cultural articulations of visual space. This disruption is brought about by space's being organized according to the modality of an emerging form of Western, technologically driven mobility which harmonizes the flow of money, the exploitation of resources, and the generation of the desire for consumption in order to perpetuate itself. Speaking like Nietzsche, one can say that a rising dominant purpose in this way seizes hold of a domain of pre-existing practices first within the European milieu and then beyond it and begins to refashion them according to its own demands.

One could, with this in mind, stress the manner in which modernity is epitomized by social transformations which initiate a radical break that dislocates the relation between Western culture, its own past, and also the wider global environment (an essential feature of which is colonialism[98]). The post-industrial subject is a witness to this process and is so precisely in virtue of having become endowed with modern eyes. Such eyes pertain to a vision as restless and malleable with regard to themselves as the burgeoning technologically driven social order of capital which emerges to frame their gaze. Modern eyes reflect the changeability of the contemporary cultural environment and replicate it by finding its presence at work in their own nervous movements and dispositions. Here is the quintessentially modern wrenching apart of tradition that Leo Strauss laments but which will not be put back together again.

What constitutes the possibility of this break with tradition? As Heidegger's articulation of the historicality of understanding and Nietzsche's view of the transformations of practices suggest, it is an essential part of modernization to facilitate and foster the continued presence of the past within itself in a manner which necessarily gives rise to feelings of contradiction—and often in ambivalent and concealed, casually ignored ways.[99] For the modern dominant modality of change to be experienced as deterritorializing, unsettling, and dispossessing, something abides within it and the self-awareness it produces which does not entirely accord with the modernizing ethos even though it is also part of it. The past, as a chain which stretches far back, does not simply lie 'behind' the modern in the form of dead material; it also lingers within and precedes it as a necessary condition of its very articulation not in the form of a return of Straussian eternal 'nature' but as a twisted, contradictory, and uncomfortable cultural inheritance. The past has not vanished to modern vision or modern sensibility generally. Rather, it haunts them precisely to the extent that it remains a constitutive condition of the historicality of modern cultural space: it abides in the remnants of customs, procedures, and habits of looking and in the feelings which these can arouse when they become material for modern reinterpretation. Modern eyes, in other words, insofar as they occupy a cultural world concatenated out of the inheritance of earlier 'things', organs, and traditions are always already permeated by the possibility of archaic visualities which exist in tension with the dominant order of the present.

This conception is already implicit one way or another in the manifestations of modernity's self-awareness discussed above. Nietzsche in the 1880s offers an account of modernity that stresses its historical retention of ancient forms and the idea that the contemporary human being is subject to primitive emotional demands and habitual judgements that betray the inheritance of his or her forbears; Simmel's conception, too, resounds with this tension. A parallel implication is also present in Baudelaire. In Baudelaire's conception, modernity is not only what has been attributed to him above, i.e. the 'transient', 'fleeting', and 'contingent'. There is in addition a counterbalancing element made manifest by what he terms the other 'half' of art. This aspect of art, Baudelaire claims, is concerned with 'the eternal and immutable'.[100] It is, in other words, bound up with those aspects of experience which are older, with what endures in the midst of modern transience. Such aspects are not at one with the 'Now' which dominates modernity's present and yet they obstinately persist within and shape it.[101] As David Harvey notes, it this 'dual formation', uniting elements that sit uncomfortably together, which so presciently captures the instability evidenced by subsequent conflicting modulations of modernist self-understanding as it shifts restlessly back and forth between the polarities of historicizing awareness and a universalizing desire (of the kind at work in Leo Strauss's thought) which draws on the sense of a past that is at once elusive and yet stubbornly present within its milieu.[102]

To take this doubled aspect of the modern seriously means rethinking the visual elements at work in the post-industrial gaze in terms of an archaic historicality that pervades and disrupts it. Understood in these terms, the terrain of contemporary visuality would not simply be passive testimony to modernizing technology's rising hegemony. It would, instead, be a historically opened up space whose very opening is continuous and erupts from the persistence within it of conflicting modes of sensibility. This persistence is hinted at in passages in the work of two writers (one situated at the opening stirrings of cultural modernity, the other at its *fin de siècle* modulation into the twentieth century) which powerfully convey the continuity of modernity's doubled awareness. The first of these is seventeenth-century thinker Thomas Hobbes. As one of the earliest proponents of modernization in the shape of the Enlightenment (the movement which was to become progressive modernity's standard-bearer), Hobbes is a self-aware and committed modern. In

Leviathan (1651), he derides anyone who cleaves to the authority of ancient learning as being dogmatically regressive. Old knowledge is today's garbage: the ancient past must be left behind since it has a damaging influence that can make it worse than useless. However, even as he dismisses what is ancient, Hobbes cannot help inserting an archaic slant into the modern standpoint he would have supplant it. This insertion pertains to a similarly decisive ambivalence as that of Baudelaire's characterization of modernity two centuries later. We, says Hobbes, owe the past in the capacity of its agedness nothing: to 'the Antiquity it self I think nothing due'. And why? The answer is put straight enough: 'For if we will reverence the Age, the Present is the Oldest'.[103] For Hobbes, being modern means embracing the rational-technological 'Present' and so being a creature that is forward-looking, methodological, scientific, and above all inclined to embrace the promise of the machine—a machine, by the way, which in the form of the Leviathan itself has eyes that allow for surveillance after the manner of an optical instrument.[104] Hobbes thus embraces thinking through the implications of the machine in the registers of politics and knowledge, and in doing so, he follows the modernizing imperative to make of thought an industrialized form.[105] Yet on the other hand and at the same time, it turns out that for Hobbes being modern also means dwelling in what are assuredly the 'Oldest' of times: when looked at from the standpoint of the present, the ancient past comes to represent something which, in being older since it came before now, is paradoxically also something far *younger* than today is. We moderns, Hobbes points out, are also the most ancient, for we are always far older in virtue of our inheritance than they who lived during the immature age of humankind could ever be.[106] The modern is not youthful at all. In its achievement of rational knowledge it is that which is 'Oldest' and hence endowed with the ancient authority of the elder.

Two hundred and fifty years later, the strange boy called 'Little Father Time' featured in Thomas Hardy's *Jude the Obscure* stands as one of the most unsettling and questioning elaborations of the continuing awareness implicit in Hobbes's blunt and less than ideally self-reflexive self-assertion. Hardy's *fin de siècle* fictional figure encapsulates modernity's distorted and contradictory temporality. Little Father Time may be a boy, but he is a boy who exudes the aura of an old man: 'Little Father Time is what they always called me. It is a nickname; because I look so aged, they say'.[107] His youth is un-youthful; he embodies the disturbing sense that modern newness (which Hardy takes special care to show the boy exemplifies, being one of 'these preternaturally old boys [who] almost always come from new countries'[108]) exists hand in hand with a precipitate and unnatural decrepitude. He is 'Age masquerading as Juvenility', the contradictory nature of the modern represented as a human face.[109] It is as if within such self-understandings the newness which in modernity is relentlessly pursued by way of the overturning of tradition is accompanied by a strange historical reversal which becomes more evident the more intense the modernity, the more *saturated* the modern cultural milieu becomes. The greater the sense of newness that surrounds and permeates us, the older we immediately come to feel. The greater the sense of the new, the more the feeling that what is modern is accompanied by the intrusion of disturbing and increasingly powerful and contradictory archaic qualities that return not merely to haunt it but to inhabit it.

4. HYPERMODERNIZING EYES

The word 'saturated' has just been used in relation to the fluidity of the modern cultural milieu. It is this saturated condition which opens out on to hypermodern vistas. Hypermodernity is liquefied or saturated modernity—modernity taken to the limit at which it begins to break down, where the chaos of image which characterizes the modern begins to go into overdrive. In brief, hypermodernization is the unsettled condition of destabilized modern life but intensified, faster, and ever more centrifugal in its effects. This saturation is also at the same time characterized by an intensification of the archaic elements which the 'now' of the modern retains within itself.

With this in mind, it is worth pondering in a fairly brief way, and by way of contrast, two contrasting accounts of the hypermodern. One is Gilles Lipovetsky's, whose view is distinctive but also problematic. It is informative in that it highlights well some important general features of the hypermodern. It is problematic insofar as it cleaves to a questionable faith in the ability of the free-market to rejuvenate social order and ignores the disruptive, archaic elements inherent in modernity outlined above. The other is the account of 'liquid modernity' suggested by Zygmunt Bauman.

Given its problematic aspects (which will become clear below) Lipovetsky's account is perhaps best treated as a useful starting point for initially broaching the hypermodern. Lipovetsky argues that hypermodernity signifies a condition in which in 'every domain there is a certain [kind of] excessiveness, one which oversteps all limits, like an excrescence'.[110] This captures well a sense of something at work in the hypermodern that is fluid but in a relatively viscous way. Being excreted suggests that the hypermodern tends to the appearance of something that has been squeezed out of the social body like an unintended waste product. Behind this excrescence lurks the pressure of contemporary technology which, as a powerful driver of change within industrial and post-industrial culture, engenders an atmosphere of relentless revision which modifies self-understanding by refashioning our cognitive propensities.[111] One might add that the advent of 'digital media' has intensified this situation in the form of what has been memorably termed 'the participatory condition' of the digital era.[112]

For Lipovetsky, hypermodern visuality tracks patterns of dissolution in the shape of the ceaseless pursuit of a 'transparency' that erodes any feelings of depth that might allow things to appear stable. This pattern is borne witness to by the generation of 'images of the body produced in the hyperrealism of porn; television and the shows it broadcasts' and the 'millions of cameras and other electronic means of surveillance and citizen identification [. . .] installed in the streets, shopping centres, public transport and business'. These features, portrayed by him in a manner reminiscent of Gilles Deleuze,[113] play out the collapse of disciplinary order in favour of 'the society of hypersurveillance'.[114] Hypermodernizing society is epitomized by a loss of traditional limits capable of governing the individual's behaviour, so surveillance steps in to police the disruptive spaces the loss of these traditional forms opens up.

This sense of loss has the knock-on effect of permeating subjective awareness, rendering it prey to becoming chaotic and contradictory also. We tend to consume with little attention to the requirements of need as opposed to desire[115] but are happy to moralize even as we do so (global warming, for example, prompts outrage and near inaction in equal measure amongst most people in the West as they fret then forget before unthinkingly taking a car journey or boarding a flight). Some of us seek 'extreme' experiences without questioning the notion that extremity is a good (consider the popularity of 'extreme sports' or the relatively casual use of drugs for enhancing 'performance' in many walks of contemporary life). Likewise, we encounter ourselves in the context of ever more ubiquitous and disturbing images which represent identity in terms that highlight the relation between the worth of a person and such things as body shape.[116]

Lipovetsky's account points to the primacy of 'presentism' ('the worship of the present'[117]) as a powerful aspect of this hypermodern obsession with the image. For him, presentism is driven by the shift from the earlier modernist economy of production to an economy of unfettered and highly visible consumption. Hypermodern society reveals itself to be gripped by a principle which promotes 'an excess of goods, images and hedonistic temptations'.[118] This tendency leads not merely to the undoing of traditional forms. It also favours 'the reign of urgency'[119] which manifests itself in a demand 'for the immediate satisfaction of needs' which in turn further renders problematic already endangered conceptions of identity based upon traditional categories (e.g. of race, 'gender, sexual orientation or historical memory'[120]).

One might say that, in Lipovetsky's account, hypermodernity threatens to become a dangerous era of cult image–worship and consumer-idolatry. But

Lipovetsky pulls back from the full implications of this. He argues, instead, that while the hypermodern condition may be one of technologically driven consumerist crises dogged by insecurity, it at the same time promises to foster an environment in which the acknowledgement of new conceptions of identity can come to the fore. In hypermodernity, 'a technocratic and market-driven spiral [. . .] is accompanied by a unanimous endorsement of the common roots of humanist and democratic values' and by a 'paradoxical [. . .] frivolity'.[121] The divided nature of the hypermodern as Lipovetsky conceptualizes it is revealed here for he holds that fetishistic presentism also contains within it a humanism that endorses feelings of moral and political optimism. With this one comes to the most problematic aspect of Lipovetsky's account. He returns his reader to a comforting faith in the idea that there remains within the terrain of even an almost entirely unfettered, consumerist culture a core democratic sensibility which, in harking back to the Enlightenment vision of a free-market liberalism capable of cultivating the ideals of scientific and social progress, engenders more humane possibilities than a more purely technologically oriented understanding of it might imply. In short, Lipovetsky attempts to reinsert the idea of the human understood as an autonomous category back into the hypermodern. For Lipovetsky, the ceaseless present of the hypermodern can hence be escaped for the 'presentism that now rules our lives [. . .] never ceases to open out on to something else'.[122] Short-termism must give way to something more enduring, an 'idea of progress' rooted in 'a hypermodern future': 'the power of the future has not been destroyed: it is simply no longer ideological and political, but borne along instead by the dynamic of technology and science'.[123] Science and technology, in short, are simply the latest vehicle of a nature that precedes them and will come to fruition by way of it. Thus, in the face of disruption brought about by technology, the only way to go seems to be to repeat the modernist gesture and embrace the humanist understanding of technological progress that underpins it ever more fiercely.

Lipovetsky's ultimately optimistic account of the hypermodern seems naive in its emphasis on presentism's redemptive potential. One can respond to hypermodernization with rather different sentiments. Against such optimism stands the fact that 'the reign of urgency' and image-worship which Lipovetsky identifies with the burgeoning of consumerist needs has revealed a potential to provoke forms of political expression that can take on starkly inhumane and authoritarian forms that seem to contradict the ambitions of liberal sentiment.[124] Hypermodern conditions, being divided, also nurture bleakly anti-pluralist sensibilities alongside other more tolerant, internationalist ones. Global communications systems facilitate the transmission of images that inflame and cause anxiety on a new scale. In the West, where the economic dominance capable of securing basic employment no longer seems assured due to competition from other players of the game of post-industrial production which 'outsources' work, blue-collar worries about the future have provoked expressions of nationalist political reaction. Beyond the West, too, 'modernization', which has generally brought with it the disruption of a sense of tradition and continuity with the past, stands revealed not only in its ability to undermine hitherto stable social arrangements but also in its impotence to compensate for the disruption caused (as is eloquently discussed by Jürgen Habermas in the context of the notorious terror attacks on the World Trade Center in New York in September 2001[125]).

This tendency of hypermodernization to engender reactive crisis has been well summed up by Zygmunt Bauman in the term 'liquid modernity'.[126] Liquefaction undoes stable relationships and generates instability. Hypermodernizing crises, as Bauman notes, tend for this reason to lead to powerful counterreactions. These reactions are intensified by the hitherto unparalleled speed of dissemination of images and words in a world both joined and divided by a common technology. Reaction against globalization is most tellingly revealed in the West by a growing desire for anti-modern forms of authority (so-called 'populism') and an accompanying nostalgia for the safety glimpsed by way of ideas of stable customs and traditions of the past.[127] This is somewhat different from an optimistic faith in a bright future secured by relentlessly progressive technical and scientific expertise.[128]

In political terms, Bauman notes, one can summarize this situation by considering the emergence in public discussion of the contention that the modern nation state as codified three centuries ago by Thomas Hobbes (a figure mentioned above as lurking ambivalently at the doorway of modernist political discourse) has failed to 'do its job properly'.[129] For Hobbes, the function of the state is to control and ultimately limit human aggression by 'civilizing' humankind.[130] The problem is that in reality the achievements of the modern era have proved to be somewhat different. What the Hobbesian and, following its pattern, the liberal-democratic state of the free-market which came to the fore within industrial modernity achieved was not an extirpation of violence in favour of the pursuit of scientific and technological perfection but violence's concealment, refinement, and intensification. In short, modernity was built upon an invisible condition of repression. With the inception of the modern state was enacted merely a

> reform of human *manners*, not human *capacities*, *predispositions and impulses*. In the course of the civilizing process, acts of human violence were shuffled out of *sight,* not out of *human nature*, as well as 'outsourced' [. . .] to professionals (bespoke tailors, so to speak, of violence), or 'subsidiarized' to lesser, 'unclean' humans—e.g. slaves, bondage demi-slaves or servants.[131]

The liberal distinction between public and private self, for example, preserves and fosters hidden forms of violation precisely because it protects what is hidden by legitimating its concealment. With the saturating world of transparent, destabilized technological space exposing hitherto protected aspects of life, this model of the state, which ironically played a central role in the contemporary liquid condition's development and colonization of global space, begins to lose its moral integrity and so unravels. The violence that was hitherto contained (i.e. repressed but not annulled) by the nation state overspills in a world that becomes increasingly obsessed with making things visible, communicable, at all costs.

Modernity, in other words, is typified by a political sphere that cultivates a visuality which, in spite of its urge to the contrary, conceals.[132] This can be tellingly linked to what Eduardo Cadava has called the modern tendency to 'transform the world into an image'.[133] When it becomes associated with the image, with mechanical 'representation', visibility gets confused with proximity, closeness, authenticity. The idea is that if the world is made entirely transparent by being everywhere illuminated, then enlightened truth must ultimately reign. However, 'the desire to abolish distance only leads to more distance', for the light something casts 'is never identical to itself', which 'means that it is always inhabited by a certain distance or darkness'.[134] Swathed in a technological light of its own making, the modern liberal state testifies to the truth of this claim by beginning to crumble.

Bauman notes the ways in which this crumbling is made manifest. The contemporary mode of liquefaction, otherwise known as 'deregulation' or 'modernization', atomizes the remnants of stabilizing social structures that persisted in modernity's diminishment of tradition and substitutes in their place an ethos of (often carefully regulated) violence at the heart of the social domain.[135] This violence is normalized by being linked to the refashioning of the public realm as a domain of transparency subject to interminable and omnipresent surveillance. Transparency, since it is meant to ensure fair competition, is assumed to be an intrinsic good. But like all forms of illumination, it is also endowed with the potential to corrode trust, belief, and hence authority. Everyone is supposed to be in competition with everyone else (competition is 'healthy') and comes to regard everyone else accordingly as an observed observer. The contemporary shaping of education is a de facto admission of this state of affairs. Aided and abetted by a combination of pseudo-scientific educational methodologies and market-orientated government interventions, education has become focused on 'skills training'.[136] Such training involves a realignment (and reduction) of the role of teaching to an activity analogous to the equipping of students with armaments and the means for defence. The skilled-up, 'kitted' individual who leaves school or university now is meant to do so not as the possessor of an enriching knowledge but as someone equipped with the tools supposed to enable them to negotiate a ruthlessly competitive environment in which all comers are envisaged as falling into the category of either vanquisher or vanquished—in which everyone is a potential enemy and so needs to be kept an eye on.

All celebrate the competitiveness of mutual rivalry and the submitting to the judgemental gaze of the spectator as the norm. In each case, the viewer overlooks combatants situated in an arena from the safe distance mediated by the desktop computer or tablet or sitting-room television screen. A distant, controlled violence becomes idle, detached amusement, its images and dramas as easily absorbing in the moment as they are forgettable in the long term. The war of all against all which Hobbes identified and from which he envisages humans retreating into the safety of social life regulated by state power now returns as a regulated public spectacle by the grace of the 'culture industry':[137] as entertainment.

According to Bauman, the authority which Hobbes sought by way of the state has evaporated and 'is nowhere in sight'.[138] This opens the way for what one might call (to coin an ugly word for an ugly phenomenon) unlimited Leviathanization. In Bauman's words, such a condition is characterized by 'the co-presence of numerous, all-too-numerous, big, small and tiny Leviathans, gravely malfunctioning'. It is this malfunctioning condition and the anxieties it provokes, Bauman notes, which triggers the nostalgic, retroactive (and indeed, thematically speaking, Hobbesian) counter-desire for an absolute protecting authority capable of engendering once more feelings of safety in a situation which comes to feel more and more like ongoing civil war. With this turn, liquid- or hypermodernization threatens to initiate a return to tribalism, to a condition of violent autochthony on a grand scale[139] as primitive sentiments that revolve around belonging and safety hook themselves onto the remnants of vanishing traditions and customs and inject them with new life by way of communications technologies.

Violence has never been far away from modernity. As the great wars of modernity that extend from the Napoleonic era to the late twentieth century reveal, it is deep within it. Indeed, it might well be said that a certain relation to violence hides within the modern as a founding condition of its existence. This being the case, it is perhaps hardly surprising that hypermodernity is prey to the same tendency.

Lipovetsky's and Bauman's contrasting views of hypermodernization or the liquid modern condition are perhaps most strikingly characterized by the way they conceptualize its relation to time and, above all, the dominant time of the present. Lipovetsky's optimistic conceptualization relies on dubious assertions concerning the future. Presentism is, on such a view, ultimately redemptive. The hypermodern present does not merely turn toward the future: it is itself already, in an important sense, the future as a projective visualization of the present in redeemed form. This view implicitly draws on the idea of the free market of capital and takes on the form, perhaps unsurprisingly, of a kind of economic surplus. It also draws on the notion of an ability to see what eyes cannot in reality penetrate: tomorrow is always foremost a projection of what was, in virtue of which what will be is projected in ghostly form. From Lipovetsky's standpoint, the future is held to be already contained in embryo within the hypermodernizing present as a benefit generated by our ongoing investment of faith in the promise of a forever-burgeoning technology. The future, in other words, takes the form of a kind of technological surplus whose redemptive wave will rise up and carry us with it over the threatening rocks of presentist instability.

In essence, this view amounts to an assertion of the primacy of an essentially visionary and prophetic conception of human will. Such a will would claim the right to see the future as its possession and celebrate itself in and as this projected futurity. Time, from such a standpoint, is something human-technological mastery can control by asserting its power over destiny through its ability to liberate the present by liquidating the past's hold over it. What is asserted in the unity of present willing and the will's destiny is a vision of a future that will overcome age-old anxieties—the fear of nature, of the uncontrolled, of pain and uncertainty. One might deem this will as being characterized by its 'Oedipal' vision for it invokes the human cunning which challenges the power of pre-ordained fate that Sophocles' Oedipus celebrates before his downfall. Here is one form of return to something archaic within the hypermodern to which it is worth returning. [140]

Unlike Lipovetsky's, Bauman's view can hardly be called optimistic. But who today needs optimists, especially when so many have a tendency to speak in terms which attest to having attended too many managerial training courses? Bauman notes how liquid modernity is marked not simply by a sense of breaking with the past but, as the notion of liquidity implies, by an unleashing of a multiplicity of uncontrollable flows and slippages which dissolve the past's constructive relation to the present. Bauman's account reveals liquid modernity to undo the space of tradition and supplant it with over-spilling excess. The traditional possibilities of projection which are taken to endow life with meaning, including the modes of visualization which allow people to situate themselves imaginatively in a stable order of things, are withdrawn as tomorrow recedes into the contemporary dissolution of social forms. The future becomes opaque and threatening; the victim of liquidity feels him- or herself to be a person darkling. We are faced with the threat of drowning in the flooding light of our own technological experimentation and its engulfing, unpredictable consequences. Understood in this way, the dynamic of modernizing liquidity is inimical to the generality of human willing since it undermines conventional frameworks which make continuity and predictability possible in life. The promise of liquid modernity does not signal the arrival of a technologically controlled world capable of giving rise to a more humane liberalism. Rather, its wave rises up threatening to return us to the archaism of a violent and primitive authoritarian past.

The movement of this wave of 'liquid modernity', the condition in which advanced technological-social forms which seek to cultivate a limitless transparency coexist alongside more opaque archaic and outmoded elements that are intrinsic to them, amounts to a movement which encapsulates the notion of 'hypermodernization'. To trace the outline of this movement further and to ponder its archaic aspects in more detail, one must turn to the thought of Paul Virilio.

NOTES

1. See David Hume, *A Treatise of Human Nature*, ed. L. A. Selby-Bigge (Oxford: Clarendon Press, 1990), sec. 1. For Hume, 'impressions' are the direct result of sensory experience and the ideas we derive from them are 'the faint images of these'.

2. David Hume, *An Enquiry Concerning Human Understanding*, in *Enquiries Concerning Human Understanding and Concerning the Principles of Morals*, ed. L. A. Selby-Bigge (Oxford: Clarendon Press, 1988), sec. 12.

3. Richard Rorty, *Philosophy and the Mirror of Nature* (Princeton, NJ: Princeton University Press, 1979), 59. When one talks about 'modern philosophy', therefore, one is often referring to the inheritance of this variety of empiricism in its various articulations.

4. Empiricism's most famous modern exponents included John Locke (1632–1704) and David Hume (1711–1776). Where one line of thought, the form of rationalism often associated with figures such as René Descartes (1596–1650) and Gottfried Leibnitz (1646–1716), argues that ideas originate in the intellectual space of the mind, the empiricist holds them to derive from the immediately given sensuous space of the external environment.

5. See Gilles Deleuze, *Empiricism and Subjectivity: An Essay on Hume's Theory of Human Nature*, trans. Constantin V. Boundas (New York: Colombia University Press, 1991).

6. John Locke, *An Essay Concerning Human Understanding*, ed. P. H. Nidditch (Oxford: Oxford University Press, 1998), chap. 1, sec. 15.

7. See John Dewey, 'The Postulate of Immediate Empiricism', in *Darwin and Philosophy and Other Essays* (New York: Prometheus, 1997), 226ff; and 'Having an Experience', in *Art and Experience* (New York: Perigree, 2009), 35ff. Another example is furnished by Gilles Deleuze's reading of Hume in *Empiricism and Subjectivity*.

8. See Michael Oakeshott, *On Human Conduct* (Oxford: Clarendon Press, 1991), 55ff.
9. Martin Heidegger, *Being and Time*, trans. John Macquarrie and Edward Robinson (Oxford: Blackwell, 2005), 89–90.
10. Heidegger, *Being and* Time, 97–98.
11. Heidegger, *Being and Time,* 97.
12. Heidegger, *Being and Time,* 98.
13. Heidegger, *Being and Time,* 99.
14. Heidegger, *Being and Time,* 153.
15. Heidegger, *Being and Time,* 154.
16. Heidegger, *Being and Time*, 154–55.
17. Vilém Flusser, *Towards a Philosophy of Photography*, trans. Anthony Mathews (London: Reaktion, 2006), 29.
18. Like Ludwig Wittgenstein, Heidegger seeks 'to recover an understanding of an agent as engaged, as embedded, in culture, a form of life, a "World" of involvements, ultimately to understand the agent as embodied' (Charles Taylor, *Philosophical Arguments* (Cambridge, MA: Harvard University Press, 1997), 61).
19. Martin Heidegger, 'The Origin of the Work of Art', in *Basic Writings*, ed. David Farrell Krell (London: Routledge, 1978), 143–212.
20. Martin Jay, *Downcast Eyes: The Denigration of Vision in Twentieth-Century French Thought* (Berkeley: University of California Press, 1994), 9.
21. Friedrich Kittler, *Optical Media: Berlin Lectures*, trans. Anthony Enns (Cambridge: Polity, 2010).
22. Knowledge of the *camera obscura*'s functional principles can be traced back to at least the times of Aristotle, whilst the first rigorous description was offered in the eleventh century by the Muslim scholar Alhazen. In its most basic form the *camera obscura* is, as the Latin name suggests, a darkened room with a small hole in one wall through which light rays enter that are projected onto a facing wall to form an inverted image (the small hole acts as a lens). Its initial uses included looking safely at solar eclipses. In the Renaissance, with the refinement of glass lens–making techniques, the *camera obscura* became an artist's primary aid for drawing landscapes and portraits (varieties were developed which were portable, although the first of these still required the viewer to sit inside them). For a usefully concise account of the history of the *camera obscura*, see Alison and Helmut Gernsheim, *A Concise History of Photography* (London: Thames & Hudson, 1971), 9–15.
23. Kittler, *Optical Media*, 49.
24. Kittler, *Optical Media*, 62.
25. Kittler, *Optical Media*, 49.
26. Kittler, *Optical Media*, 50.
27. Kittler, *Optical Media*, 50.
28. See note 22.
29. Kittler, *Optical Media*, 58.
30. The practitioners of pointillism, Kittler notes, constructed their images out of coloured dots, i.e. codes, which were developed from the technological development

of the colour photographic image and the dots and dashes of telegraph codes (*Optical Media*, 205).

31. Walter Benjamin, 'The Work of Art in the Age of Mechanical Reproduction', in *Illuminations*, ed. Hannah Arendt, trans. Harry Zorn (London: Pimlico, 1999), 218.

32. Benjamin, 'The Work of Art', 227, 230.

33. Günther Anders, 'On Promethean Shame', in Christopher John Müller, *Prometheanism: Technology, Digital Culture and Human Obsolescence* (London: Rowman & Littlefield, 2016), 74, 76. Patterns of ancient orgiastic ritual, Anders notes, appear in domains which this tendency of the machine gets a hold of: in the dance club, for instance, where mechanically generated rhythms draw bodies into the ecstasy of 'an *Industrial Dionysus-Cult*' (78). The dance hall of Anders's times is that of the jazz night club. As Christopher Müller notes, Anders's attitude here 'replicates similar prejudices of a number of émigré German thinkers' who found themselves in the United States in the period of World War II, most conspicuously Adorno (77). One can, of course, appreciate jazz or other popular forms of music whilst nevertheless appreciating Anders's point also.

34. This is not an issue of self-awareness but of understanding and hence also self-understanding. 'Being historical' does not necessarily mean thinking of oneself as historical.

35. Heidegger, *Being and Time*, 41.

36. See Jacques Derrida, *Of Grammatology*, trans. Gayatri Chakravorty Spivak (Baltimore, MD: Johns Hopkins University Press, 1976), 70.

37. Looking and comprehending are conjoined. Consider the Old English '*seon*', 'to look', but also 'to understand', 'to perceive', or 'to experience' something. The proto-Indo-European '*sekw-*', 'to follow', is the possible source of this—'to see' being to follow something with the eyes.

38. Jonathan Crary, *Techniques of the Observer: On Vision and Modernity in the Nineteenth Century* (Cambridge, MA: MIT Press, 1992), 1.

39. See part 2 of this volume.

40. Strauss is one of the key figures associated with the rise of 'neo-conservatism'. See Leo Strauss, 'The Three Waves of Modernity', in *Political Philosophy: Six Essays by Leo Strauss*, ed. Hilail Gilden (Indianapolis, IN: Bobbs-Merrill, 1975), 81–98. For Strauss, modernity, when viewed aright, is exemplified by a corrosive historicism. The corrosion comes along in what Strauss calls three phases, or 'waves'. The first wave envelops European thought in the fifteenth and sixteenth centuries with Niccolò Machiavelli and Thomas Hobbes, who first sever the link between the spheres of natural law and politics (Machiavelli) and then reassert natural law in destructive form as the unlimited right to self-preservation in a state of nature consisting of pure violence (Hobbes). The second wave is initiated in the eighteenth century by Jean-Jacques Rousseau, who radicalizes the Hobbesian state of nature by inverting it to portray humankind as moving from a condition of natural innocence to an oppressive, civilized, social existence. The second wave of modernity thereby fulfils the first's tentative unpicking of the link forged by classical Greek thought between law, civilization, virtue, and nature by severing it altogether. The third historicizing

wave, says Strauss, is the nineteenth century's. Above all, it is Friedrich Nietzsche's. Nietzsche seizes hold of historicism and turns it into a disruptive political tool for undermining all ethical certainties and supplanting them with a philosophy of power. For the historicist, all values are cultural not natural. There is therefore no enduring and independent 'natural law' in which to anchor our moral beliefs. On Strauss's reading, modernity is, for this reason, morally implosive, for it threatens continually to collapse in on itself. Strauss opposes this with the archaic, understood as signifying a domain of universals which ensure a stable and enduring moral perspective. What is archaic, for him, is nature. Nature precedes us, and just as art copies the nature it seeks to represent, so in morality and politics nature always provides the models we ought to emulate in the form of ancient tradition. Thus the 'oldest things', enduring and eternal, precede and ground human history itself. Strauss's view presupposes that when the authentic past erupts in the present, it does so in a form that does not change over time. It also presupposes a firm link and causal order between abstract concept and visuality. First principles are lasting, timeless concepts endowed with a sense that steadfastly points toward a specific moral purpose—the 'good'. Strauss assumes that practices and traditions are guided by the kind of necessity that is often assumed to be a defining characteristic of concepts. He thinks concepts (abstract 'ideas') are stable and enduring, whereas historical practices are unstable and temporary. This notion grounds his Platonic view that behind history lie unchanging, primitive conceptual forms whose meanings are fixed. What is older (i.e. ancient knowledge), is supposed to be more authentic *because* it is enduringly old and hence closer to our origin in the condition of nature. Strauss holds modern thought, with its turn to historicizing, to be a move away from this origin, spurning the ancient affirmation of a timeless natural morality and locating the source of judgement instead in the contingencies of the human-made cultural domain of time and place.

41. Friedrich Nietzsche, *On the Genealogy of Morality*, ed. Keith Ansell Pearson, trans. Carol Diethe (Cambridge: Cambridge University Press, 2017), II, sec. 12–13.

42. See John Dewey, *Darwin and Philosophy and other Essays* (New York: Prometheus, 1997), 1ff.

43. Nature, Nietzsche points out, squanders (Friedrich Nietzsche, *The Gay Science*, trans. Walter Kaufmann (New York: Vintage, 1974), sec. 1).

44. Likewise, truth can be conceived as arising from error. We talk of the environment in terms of concepts which we pragmatically assume to refer to something that corresponds to them. Truth, taken this way, comes to be understood as correspondence between concept and thing. The origins of our concepts, however, are not very likely to reside in any principle of correspondence. That would, for one thing, presuppose the relation of correspondence between concept and thing to precede the existence of the concept which is part of the relation, which makes no sense. One can, in contrast, think of concept use as the pragmatic and customary application of labels which allow the environment to be negotiated rather than 'represented'. Concepts are not, it follows, 'representations' in the picturing sense presupposed by the notion of correspondence. Truth, equally, cannot be the objective correspondence between representations, their concepts, and the things in the environment. Yet we talk as if this were the case and for the most part get by perfectly well on the basis of doing

so. Truth talk, it follows, arises from error: it emerges from its 'opposite'. See in this connection, Friedrich Nietzsche, *Beyond Good and Evil*, trans. R. J. Hollingdale (London: Penguin, 2003), sec. 1.

45. Nietzsche, *On the Genealogy of Morality*, II, sec. 12.

46. See *The Gay Science* section 1 for Nietzsche's view of this. Humankind is the 'fantastic animal' in that it needs meaning, but no meaning proffered ever lasts for long. The need for meaning gives rise to what Nietzsche calls 'the age of tragedies'. To perceive this need for what it is, i.e. as a mad kind of compulsion, opens up the perspective of the 'eternal comedy' of human existence.

47. See Nietzsche, *On the Genealogy of Morality*, I. Etymology is the 'pointer in the *right* direction as to what the terms for "good", as used in different languages, mean' (I, sec. 4). Moral talk, Nietzsche argues, has its origins in the powerful social groups of the ancient world, the masters or nobles. The originally noble practice of naming, 'possibly the origin of language itself' (I, sec. 3), expresses the power of the namer over the named (be it a named thing or person). Values testify to self-assertion: originally, the 'good' are the powerful themselves, who name and thereby evaluate their world accordingly, whereas those whom the nobles deem 'bad' (i.e. ignoble, common, plebeian, and unworthy) are simply the powerless slave members of the lower orders (I, sec. 5). Meanings shift, however. The noble evaluative practice of coining values is subsequently taken up by the subjugated orders (Nietzsche deems this usurpation to mark the moment of the birth of 'slave morality') who adopt the term 'good' and reverse its meaning so that it celebrates the dispossessed themselves: 'good' now denotes the powerless, the meek, the oppressed, those who can do no harm. By the same token, the subjugated coin the contrasting term 'evil' to denote their oppressors—'evil' persons being the noble originators of the evaluative practice which the slave orders subsequently seize and turn against their oppressors, adopting and simultaneously re-evaluating the notion of the 'good' in the process. The point is that the linguistic practice of naming and valuing has its origins in one set of power relations, and the initial meanings associated with evaluative terms reflect this. The slave's usurpation of noble evaluative practice is marked by a radical reinterpretation of the meanings made available within the practice of naming itself. With the rising dominance of slave moral discourse, different meanings which all but obscure the origins of both the evaluative terms and the practice itself come to the fore.

48. For a detailed discussion of this, see Peter R. Sedgwick, *Nietzsche's Economy: Modernity, Normativity and Futurity* (Basingstoke: Palgrave, 2007).

49. These measures are punishments. In the primitive community the most effective punishments utilize painful means—pain is easily remembered. Such means, Nietzsche argues, constitute the basic '*technique of mnemonics*' of our proto-human ancestors. Any action perceived as being detrimental to the community was punished by the imposition of pain since, Nietzsche speculates, pain is the most effective stimulus to memory. The effect of this, in the long term, was the creation of the human animal's moral memory, which associated breaking the bonds of communal life with painful things. Here, Nietzsche adds, lie the origins of the psychology of guilt which is so prominent a feature of human cultures (*On the Genealogy of Morality*, sec. 3).

50. Nietzsche, *On the Genealogy of Morality*, II, sec. 5.

51. This is so to such an extent, Nietzsche contends, that the setting of prices, working out of equivalents, and the like 'in a certain sense *constitutes* thought' (*On the Genealogy of Morality*, II, sec. 8).

52. Exchange ultimately gets transferred from the basically functional relation that inheres between biological individuals and the community to the realm of relationships between persons in the form of the creditor-debtor relationship. This, says Nietzsche, is the most ancient and 'most primitive personal relationship there is [. . .] here person met person for the first time, and *measured himself* person against person' (*On the Genealogy of Morality*, II, sec. 8).

53. The human being came in this way to regard itself as 'the calculating animal as such', took the 'germinating feeling' associated with notions of duty, compensation, duty, and the like and merely '*transferred*' them 'from the most rudimentary form of the legal rights of persons to the most crude and elementary social units [. . .] together with the habit of comparing power with power, of measuring, of calculating' (Nietzsche, *On the Genealogy of Morality*, II, sec. 8.).

54. See, for example, Nietzsche, *Beyond Good and Evil*, sec. 20. The argument here is that language pertains to a structure of inheritance which guides and fashions habits of thought. Thus the 'family resemblance' of the philosophizing of Indians, Greeks, and Germans is explained by their common grammatical inheritance.

55. Nietzsche is keen to hold that habitual adherence to practices is a condition of abstract conceptual forms. To take one example: the words 'pure' and 'impure' were originally 'signs of different estates'. Terms of this kind pertained to little or no symbolic sense: 'all ancient man's concepts were originally understood [. . .] as crude, coarse, detached, and in particular *unsymbolic*. From the outset the "pure man" was just a man who washed, avoided certain foods which caused skin complaints, did not sleep with the filthy women from the lower orders and had a horror of blood—nothing more, not much more!' (*On the Genealogy of Morality*, II, sec. 6).

56. The view of Nietzsche offered here has much in common with that of Friedrich Kittler. 'In the second essay of *On the Genealogy of Morality*, knowledge, speech, and virtuous action are no longer inborn attributes of Man'. The form of expression called 'writing' hence ceases to be 'a natural extension of humans who bring forth their voice, soul, individuality' by way of it. Rather, the norms, habits, and customs that we find ourselves observing become explications of us. See Friedrich Kittler, *Gramophone, Film, Typewriter*, trans. Geoffrey Winthrop-Young and Michael Wutz (Stanford, CA: Stanford University Press, 1999), 201.

57. The origin of the word for 'human', Nietzsche notes, expresses this burgeoning condition of primitive self-awareness: *manas* (man) means one who measures, 'the calculating animal as such' (*On the Genealogy of Morality*, II, sec. 8).

58. Nietzsche, *On the Genealogy of Morality*, II, sec. 8.

59. An ancient numbering linked to the practice of counting sheep. For an exploration of its beguiling dramatic and musical possibilities, see the opera by Harrison Birtwistle, with libretto by Tony Harrison, *Yan, Tan, Tethera: A Mechanical Pastoral*, Universal Edition: www.universaledition.com/themes/Frontend/UniversalEdition/frontend/_public/src/js/pdfjs/web/viewer.html?file=/resources/ansichtspartituren/ue17684_pa_a3.pdf.

60. See, for example, John Berger, 'Rodin and Sexual Domination', in *About Looking* (London: Bloomsbury, 2005), 184ff.

61. John Berger, *Ways of Seeing* (London: Penguin, 2008), 45–46.

62. For the piece which first develops the idea of the male gaze, see Laura Mulvey's paper 'Visual Pleasure and Narrative Cinema', in *Film Theory and Criticism*, ed. Leo Braudy and Marshall Cohen (Oxford: Oxford: University Press, 1999), 833–44.

63. Berger, *Ways of Seeing*, 63.

64. Mulvey, 'Visual Pleasure and Narrative Cinema', 836. Such a claim is perhaps problematic in that it invokes the notion of natural propensity innocent of culture.

65. Thus 'the male gaze is also the female gaze, insofar as 'women look at themselves through the male gaze' (Laura Mulvey, in Roberta Sassatelli, 'Interview with Laura Mulvey: Gender, Gaze and Technology in Film Culture', *Theory, Culture and Society* 28, no. 5:127.

66. Foucault's account is developed from an encounter with Jeremy Bentham's *Panopticon*. See Jeremy Bentham, *The Panopticon Writings*, ed. Miran Božovič (London: Verso, 1995).

67. Michel Foucault, *Discipline and Punish*, trans. Alan Sheridan (New York: Vintage, 1977), 220, 201.

68. Mulvey, 'Visual Pleasure and Narrative Cinema', 834.

69. Hélène Cixous, 'Sorties', in *The Cixous Reader*, ed. Susan Sellers (London: Routledge, 1984), 37. Cixous develops her account of this structure by way of the notion of 'logocentrism', a term coined by Jacques Derrida to signify the 'metaphysics of phonetic writing' (*Of Grammatology*, 3). This metaphysics, Derrida holds, 'was fundamentally [. . .] nothing but the most original and powerful ethnocentrism'. It finds expression in 'heliocentrism', the metaphorical privileging of 'light' over 'dark', where 'light' is understood as being coterminous with 'truth' (27). In virtue of the power of this structuring metaphor, the kinds of force associated with embodied, tactile experience are spurned in favour of the power of the representational image. The result of this is that intensity of thought—its embodied aspect, the realm of everyday practices—is always held to be inferior to the 'knowing', commanding, and abstract picturing practice of 'pure' mental representation (so-called 'epistemology' which seeks to picture the world as a totality from afar).

70. Cixous, 'Sorties', 39.

71. Cixous, 'Sorties', 40.

72. Nietzsche, *On the Genealogy of Morality*, II, sec. 12.

73. Friedrich Nietzsche, *Thus Spoke Zarathustra*, trans. Walter Kaufmann, in *The Portable Nietzsche* (London: Chatto & Windus, 1971), sec. 2.

74. In claiming this, the reading offered here must find itself in disagreement with Alexander Nehamas's claim that Nietzsche does not accept the notion of modernity. See Alexander Nehamas, 'Nietzsche, Modernity, Aestheticism', in *The Cambridge Companion to Nietzsche*, ed. Bernd Magnus and Kathleen M. Higgins (Cambridge: Cambridge University Press, 1996). One can contrast this with the views of Robert B. Pippin or Paul S. Loeb. Both regard Nietzsche as reacting against the declining modernity he diagnoses. See Robert B. Pippin, 'Nietzsche's Alleged Farewell: The

Premodern, Modern, and Post-modern Nietzsche', in *The Cambridge Companion to Nietzsche*, 252–80; and Paul S. Loeb *The Death of Nietzsche's Zarathustra* (Cambridge: Cambridge University Press, 2010).

75. Friedrich Nietzsche, *Daybreak*, trans. R. J. Hollingdale (Cambridge: Cambridge University Press, 1982), sec. 99.

76. For a discussion of Nietzsche's relation to Darwin see John Richardson, *Nietzsche's New Darwinism* (New York: Oxford University Press, 2004). Richardson argues that Nietzsche's apparent rejection of Darwinian thought conceals a deeper engagement with its implications. Nietzsche's overt rejection, in short, is symptomatic of a deep and careful elaboration of its implications. My own view, following Eugen Fink, is that, like most things in Nietzsche's writings, the Darwinian aspects of Nietzsche's textuality often turn on its metaphorical and rhetorical implications. Nietzsche is no methodological fetishist. His 'naturalism', in short, is not to be taken as a straightforward replication and application of scientific-descriptive discourse. See Eugen Fink, 'Nietzsche's New Experience of the World', in *Nietzsche's New Seas*, ed. Michael Allen Gillespie and Tracy B. Strong (Chicago: University of Chicago Press, 1988), 203–19. For a discussion of Nietzsche's 'naturalism' see also Peter R. Sedgwick, 'Hyperbolic Naturalism: Nietzsche, Ethics and Sovereign Power', *Journal of Nietzsche Studies* 47, no. 1 (Spring 2016): 141–66.

77. Kittler, *Gramophone, Film, Typewriter*, 39.

78. Mark Hansen, *Embodying Technesis: Technology Beyond Writing* (Ann Arbor: University of Michigan Press, 2000), 2–3.

79. Marshall McLuhan, *Understanding Media: The Extensions of Man* (Cambridge, MA: MIT Press, 1994), 9.

80. McLuhan, *Understanding Media*, 18.

81. McLuhan, *Understanding Media*, 8.

82. McLuhan, *Understanding Media*, 9.

83. McLuhan, *Understanding Media*, 110ff.

84. See Karl Marx, *Capital: A New Abridgement*, ed. David McLellan, trans. Samuel Moore and Edward Aveling (Oxford: Oxford University Press, 1995), part 4, chap. 14, sec. 3:209ff. Machinery, Marx argues, is the necessary condition for the transformation of production into 'a complete technical system on its own foundation' (214). This system develops in the labourer 'new powers that are by nature fitted only for limited and special functions' (214–15). A timepiece like a watch serves as an ideal example of the first kind of 'genuine manufacture' of 'a structure produced by modern mechanical industry' (210), for it is the outcome 'of an immense number of detail labourers'.

85. Crary, *Techniques of the Observer*, 128.

86. Siegfried Kracauer, *Theory of Film: The Redemption of Physical* Reality (New York: Oxford University Press, 1960), 8–9.

87. Siegfried Kracauer, *The Mass Ornament: Weimar Essays*, trans. and ed. Thomas Levin (Cambridge, MA: Harvard University Press, 1995), 249.

88. See Walter Benjamin, 'The Work of Art', 211–44.

89. Thomas De Quincey, 'The English Mail-Coach, in *Confessions of an English Opium Eater and Other Writings*, ed. Robert Morrison (Oxford: Oxford University Press, 2013), 183.

90. De Quincey, 'The English Mail-Coach', 184, 187.

91. Jürgen Habermas, 'Modernity's Consciousness of Time', in *The Philosophical Discourse of Modernity*, trans. Frederick Lawrence (Cambridge: Polity Press, 1987), 9.

92. Charles Baudelaire, 'The Painter of Modern Life', in *Selected Writings on Art and Literature*, trans. P. E. Charvet (London: Penguin, 2006), 403.

93. Georg Simmel, 'The Metropolis and Mental Life', in *The Sociology of Georg Simmel*, ed. Kurt H. Wolff, trans. H. H. Gerth and C. Wright Mills (Glencoe, IL: Free Press, 1950), 409–10.

94. See George Grosz, 'Kannst du radfahren? [Can you ride a bicycle?]', *Neue Jugend* (June 1917): 1. For a useful discussion of this, see Kurt Beals, 'Text and the City: George Grosz, *Neue Jugend*, and the Political Power of the Popular Media', *Dada/Surrealism* 19, no.1 (2013).

95. It is worth noting that over time, varieties of *camera obscura* were developed which were portable, the first of which still required the viewer to sit inside it (see Alison and Gernsheim, *A Concise History of Photography*, 9ff). Later variants included portable, table-top versions with ground-glass screens onto which the image was projected. In essence, the principle of mobility which Crary identifies with the advent of the nineteenth-century stereoscope is implicit in these designs. This is borne witness to by the fact that these later instruments played an important role in the development of early photography. Interestingly, the design principle remains more or less unchanged even with the advent of digital imaging: the camera in a cell phone is in essence a portable *camera obscura*, a chamber with an inlet through which light is projected to an opposing wall on which is mounted a small, light-sensitive chip harnessed to a computer.

96. Crary, *Techniques of the Observer*, 145.

97. Crary, *Techniques of the Observer*, 149. Crary, drawing on the terminology of Deleuze and Guattari (see Gilles Deleuze and Félix Guattari, *Anti-Oedipus: Capitalism and Schizophrenia*, trans. Robert Hurley, Mark Seem, and Helen R. Lane (Minneapolis: University of Minnesota Press, 2000)), argues persuasively that the modern era marks the rise of a powerful hegemonic tendency in the West to overrun and decode cultural realms of all kinds, both within the European milieu and the non-European cultural and geographic terrains it is already engaged in colonizing. Relatively stable, coded domains of traditional social orders are subjected to the effects of disruptive and overwhelming flows of desire organized around capitalist production and consumption: 'Capitalism tends toward a threshold of decoding that will destroy the socius in order to make it a body without organs and unleash the flows of desire on this body as a deterritorialized field' (Deleuze and Guattari, *Anti-Oedipus*, 33). This decoding is extended by economic-industrial means and in conjunction with the use of the enhanced military power that such means facilitates.

98. For an excellent exploration of the relationship between modernity and colonialism, see Couze Venn, *Occidentalism: Modernity and Subjectivity* (London: Sage Publications, 2000).

99. See Paul Connerton, *How Modernity Forgets* (Cambridge: Cambridge University Press, 2009). Forgetting, Connerton notes, occurs in many ways. One obvious example in modern societies is the production and movement of goods and the anonymous provision of services: this means one forgets the people who are responsible

for them, who are always unknown others (142). Likewise, one forgets things simply by becoming so used to them they vanish from thought—motorways being a prime example (110).

100. Baudelaire, 'The Painter of Modern Life', 403.

101. As Jürgen Habermas comments, with this duality 'eternal beauty shows itself in the guise of the costume of the times' ('Modernity's Consciousness of Time', 9).

102. David Harvey, *The Condition of Postmodernity* (Oxford: Blackwell, 1992), 10.

103. Thomas Hobbes, *Leviathan*, ed. Richard Tuck (Cambridge: Cambridge University Press, 1994), 490.

104. Hobbes's conception of sovereign power deploys the notion of the machine as its central metaphor. The Leviathan is 'an Artificiall Man; though of greater stature and strength than the Naturall, for whose protection and defence it was intended' (*Leviathan*, 9). It is endowed with joints, tendons and nerves, a voice, hands, an ear and an eye (167–69). Indeed, the Leviathan possesses more than one kind of eye since 'Forts, Garrisons, Guns [. . .] and continuall Spyes', along with ministers sent on foreign missions, are also the Leviathan's eyes (90). Sovereign power, in other words, depends upon 'continuall' and vigilant surveillance.

105. For a discussion of this see Peter R. Sedgwick, 'Hobbes, Sovereign Power and Money', in *Credo Credit Crisis: Speculations on Faith and Money*, ed. Laurent Milesi, Aidan Tynan, and Christopher John Müller (London: Rowman & Littlefield, 2017).

106. To which it should be added that Hobbes does not by any means seek to escape from this paradox. Not far short of half of the text is devoted to an elaboration of the 'Kingdome of God' and the ancient prophecies of the Bible.

107. Thomas Hardy, *Jude the Obscure*, ed. Patricia Ingham (Oxford: Oxford University Press, 2002), 270.

108. Hardy, *Jude the Obscure*, 270. The boy has recently arrived from a 'new' (i.e. recently colonized) world: Australia. Little Father Time's later apologetic commentary on his own inability to take joy from life confirms his self-awareness of his condition: '"I am very, very sorry, Father and Mother," he said. "But please don't mind!—I can't help it. I should like the flowers very very much, if I didn't keep on thinking they'd be all withered in a few days!"' (286). Little Father Time is also the harbinger of a forthcoming cult of death, as is revealed following the child's horrific suicidal murder of his half siblings. His condition is diagnosed in the novel as being symptomatic 'of the coming universal wish not to live' (326).

109. Hardy, *Jude the Obscure*, 266.

110. Gilles Lipovetsky, *Hypermodern Times*, trans. Andrew Brown (Cambridge: Polity, 2005), 32.

111. For one take on this, see Jonathan Crary, *24/7: Late Capitalism and the Ends of Sleep* (London: Verso, 2014). A useful discussion of the cultural impact of contemporary technology on attention is offered by Claudio Celis Bueno in *Attention Economy: Labour, Time and Power in Cognitive Capitalism* (London: Rowman & Littlefield, 2017). For a prescient account of the consequences of pursuing 'the speed-life', see Arthur Kroker, *The Possessed Individual: Technology and Postmodernity* (London: Macmillan, 1992).

112. For a thorough overview of this, see Darin Barney, Gabriella Coleman, Christine Ross, Jonathan Sterne, and Tamar Tembeck, eds., *The Participatory Condition in the Digital Age* (Minneapolis: University of Minnesota Press, 2016). For a thought-provoking account of the effects of digitalization, see also Bernard Stiegler, *States of Shock: Stupidity and Knowledge in the 21st Century*, trans. Daniel Ross (Oxford: Polity, 2014).

113. See Gilles Deleuze, 'Postscript on Societies of Control', in *Negotiations, 1972–1990*, trans. Martin Joughin (New York: Columbia University Press, 1995), 177–82.

114. Lipovetsky, *Hypermodern Times*, 32.

115. Contemporary life, it seems almost needless to point out, has a problem with excessive consumption. The most obviously discussed symptom of this relates to the worldwide spread of obesity. Some thoughtful explorations of this are offered by submissions to the academic publication *Fat Studies: An Interdisciplinary Journal of Body Weight and Society*. Rachel Fox, for example, has offered a recent critical engagement that explores the relation between the dominant, 'commonsensical' idea of progress and representations of dieting which deploy the temporality of before/after comparisons by way of photographs. Such a model, as she notes, denigrates the past and present and glorifies the future ('Against Progress: Understanding and Resisting the Temporality of Transformational Weight-Loss Narratives', *Fat Studies* 7, no. 2 (2018): 216–26).

116. See n115. One of the most insightful of engagements with body image is offered by Julia Kristeva in *Powers of Horror: An Essay on Abjection*, trans. Leon S. Roudiez (New York: Columbia University Press, 1984). For Kristeva, the abject is a concealed threat to bodily self-image that threatens to overrun and disrupt bodily integrity (cancerous growths, bodily fluids, etc.). For one approach to questions of body image in relation to contemporary media, see Marika Tiggemann and Isabella Barbato, '"You Look Great!"—The Effect of Viewing Appearance-Related Instagram Comments on Women's Body-Image', *Body Image* 27 (2018). See in this connection also Maggie Wilkes and Barry Gunter, *The Media and Bodily Image* (London: Sage Publications, 2004). An interesting question arises as to whether Kristeva's notion of the 'abject' might be applied to media forms which threaten to overturn bodily integrity, i.e. as to whether unrestrained flows of digital information are instances of the abject.

117. Lipovetsky, *Hypermodern Times*, 36.

118. Lipovetsky, *Hypermodern Times*, 37.

119. Lipovetsky, *Hypermodern Times*, 50.

120. Lipovetsky, *Hypermodern Times*, 65.

121. Lipovetsky, *Hypermodern Times*, 68, 40. This includes such things as the rise in popular demand for the recognition of wrongs for national offences, the elevation of the notion of the victim, the rise of more self-critical attitudes (66), and the 'remobilization of traditional beliefs' in service of a more hybridized conception of individual identity (67).

122. Lipovetsky, *Hypermodern Times*, 41.

123. Lipovetsky, *Hypermodern Times*, 42, 43.

124. One thinks not only of the disastrous aftermath of the 2003 American-British invasion of Iraq, the repression which followed the so-called Arab Spring begun in

late 2010, or the anti-globalization movements which have sprung up in Europe and the United States over the past five years. For insightful discussions of such things, one can refer to nothing better than the essays of Zygmunt Bauman in *A Chronicle of Crisis: 2011–2016* (London: Social Europe Edition, 2017).

125. Jürgen Habermas, 'Faith and Knowledge', in *The Future of Human Nature*, trans. Max Pensky, Hella Beister, and William Rehg (Cambridge: Polity, 2003). Modernization, on Habermas's reading, takes the form of 'an accelerating and radically uprooting' (102) force which overturns established cultural norms. In the West, he notes, this onslaught of '*creative* destruction', to recall a phrase coined by Schumpeter (see Joseph A. Schumpeter, *Capitalism, Socialism, and Democracy* (London: Routledge, 2010), 73), was accompanied by a compensatory rise in living standards which was largely absent in the home countries of the assailants on the towers. Modernization is hence always amenable to taking on the appearance of destructive violation.

126. For useful discussions of some of these aspects, see Zygmunt Bauman, *Liquid Times* (Cambridge: Polity, 2007), chap. 1:5–26, and chap. 2:27–54.

127. Bauman cites in this context Svetlana Boym, *The Future of Nostalgia* (New York: Basic Books, 2001). For some discussion of this text, see Zygmunt Bauman, *Retrotopia* (Cambridge: Polity, 2017), 2–3.

128. A moment in the UK European Union referendum debate of 2016 captured the spirit of this reaction well in the form of 'Vote Leave' campaigner Michael Gove's assertion that 'people in this country have had enough of experts' (*Financial Times*, 3 June 2016).

129. Bauman, *Retrotopia*, 13.

130. See Thomas Hobbes, *Leviathan*. Despite its absolutist tendencies, Hobbes's theory is not infrequently associated with the emergence of liberalism since it codifies the political realm as emerging from the autonomous actions of individuals in a realm governed by rudimentary natural law. Hobbes's conception of sovereign power centres on the construction of an 'Artificiall Soul' which gives 'life and motion to the whole [social] body' (9) by endowing it with the unity of political will. This construction emanates from the notorious 'state of nature', a condition of civil breakdown in which everyone is at war with everyone else. This condition is overcome by the establishment of a political commonwealth and a form of sovereign power. The absolute authority of sovereignty springs from its ability to ward off civil strife. In effect, Hobbes proposes that governance involves the imposition of unity on multiplicity: the subjection of the individual to the unifying power of sovereign rule is what constitutes him or her as a citizen or political subject.

131. Bauman, *Retrotopia*, 14. Such violence, Bauman notes, is also at times sanctified. Witness the love of foxhunting or other forms of hunting in the United Kingdom; witness, likewise, the American love affair with the gun and the gun-shop.

132. For a recent polemical discussion of this tendency, see Byung Chul-Han, *The Transparency Society*, trans. Erik Butler (Stanford, CA: Stanford University Press, 2015).

133. Eduardo Cadava, *Words of Light: Theses on the Photography of History* (Princeton, NJ: Princeton University Press, 1997), xxiii.

134. Cadava, *Words of Light*, xxvi, 30.

135. Bauman, *Retrotopia*, 44ff.

136. Recent developments in education policy stress the role of technology and 'skills' in place of the discipline of knowledge-acquisition. See in this connection Daisy Christodolou, *Seven Myths About Education* (London: Routledge, 2014). The central presupposition at work in this approach, Christodolou has noted, is that in a fluid world economy driven by technological advance, 'an education that requires you to learn specific stuff will doom you to obsolescence and irrelevance. Instead, education should focus on the acquisition of skills, preferably transferable skills, which will allow the individual to adapt quickly to the inevitable change that the twenty-first century will bring' (49). In effect, traditional pupil or student, understood as a recipient of knowledge, must become a mobile 'learner'. The idea is that people 'unburdened by knowledge' (51) will be free to respond to the changing demands of an ever more fluid economic environment if they are instead equipped with 'essential skills'. The problem with such an approach, as Christodolou shows, is that there is nothing especially contemporary in the notion of 'skills' and 'problem solving' advocated by the contemporary education theorists who favour this conception. In fact, as Christodolou argues, 'old ideas' that have often been 'discredited are [simply] being warmed over and presented as being at the cutting-edge' (56). Stripped of the knowledge-context in virtue of which they are endowed with meaning, such skills become emptied of content with 'pernicious' consequences (53). One of these is diminution of the ability for what Paul Virilio has called 'mnesic choices', that is, choices which draw upon living memory (*The Vision Machine*, trans. Julie Rose (Bloomington: Indiana University Press, 1994), 13). This diminution accompanies a 'dependence on the lens' which creates an amnesiac self. The presupposition driving the cultivation of this amnesia is that given the pervasive nature of information-retaining technology in modern social orders, individual memory no longer needs to be cultivated since it is redundant. A person who has been habituated to the requisite 'skill-set' can simply look up the knowledge they want on the Internet and then apply their skills training to attain a desired outcome. The political implications of this attitude invite careful reflection.

137. The term, of course, is Max Horkheimer and Theodore Adorno's (*Dialectic of Enlightenment*, ed. Gunzelin Schmid Noerr, trans. Edmund Jephcott (Stanford, CA: Stanford University Press, 2002), 94ff).

138. Bauman, *Retrotopia*, 48.

139. Bauman, *Retrotopia*, 49.

140. For a discussion of *Oedipus Tyrannus* see part 2, chapter 4 of this volume.

Part 1

CONFLICT IN SIGHT: THE BUNKER, HYPERMODERNIZING VISION, AND ARCHAIC DEFENSIVE SPACE (APOLLONIAN, DIONYSIAN, AND SOCRATIC)

Littoral

Outside, time's tide
was petrified; saltbush withered
in the swell. Inside,

switchback surges
as a dead man wrestled darkness
in a quartz-slashed tomb.

You watched the skull.
Unseen, a winding sheet
was folded tenderly, a body

cast adrift then donned again.
Crabs inched back
to wrack. You caught him

struggling from the bunker,
shedding shadows, trying
to find his bearings by the sun

A figure stands on a decaying defensive bunker. Both figure and building are facing the sea, a thin line of luminescence edging into dark in the distance indicating where water and sky meet. The building's rounded edges resonate with the weathered stones surrounding it. The stones are absorbing the structure, which has begun to yaw and sink into them like a holed ship on water. The stones acquire a powerful kind of fluidity from this, each a molecule endowed with the suggestion of an imperceptible but powerful shared motion which is both pulling the bunker downward and threatening to wash over it.

The figure seems glued to the spot. This engenders a dominant sense of the moment captured in the image since a moment necessarily revolves around a subject and his or her world. At the same time, however, the figure's presence in its current position is undermined by adjacent ghostly outlines which impart a sense of the existence of other moments than the 'current' dominant time of the image (whether of moments before or after it is impossible to tell). This is paralleled by the blurring movement of clouds, which contrast in a similar vein with the relative stability of the slowly sinking structure and the engulfing stones surrounding it.

Everything is frozen, as one would expect with a photograph. Yet at the same time, varieties of different speeds and intensities are simultaneously suggested as being at work in the formation of the image. There is the constant speed and intensity of the radiating light emanating from the sun, which is made visible only as a pure, featureless whiteness. One does not 'see' this light, only the trace of its overwhelming intensity.[1] *Then there are the variable intensities of sky, sea, and clouds, the latter a blurring of movement, the other two simply varieties of shade and smoothed-out edges. There are also pent-up forces suggested by the stones. These simply lie there, thrown up by the combined forces of earth, air, and water. There is the explosive human threat of a projectile's speeding violence inherent in the defensive structure and the figure in silhouette, intense in its fragility. This endows a paradoxical sense of the archaic emanating from the bunker's very modernity. It is a modern structure which nevertheless recalls ancient forts and city walls. There is something primitive here. The greatest stillness and most ambivalent intensity emanates from the holes in the bunker's side which lead into an unremitting darkness—a swallowing or withdrawal of light that is no less featureless than the sun's overwhelming after-trail of pure white.*

The bunker dominates not only the foreground but the terrain behind it. It juts out from the environment, scarring it and, in this scarring, framing the figure adorning it. It opens out space by enclosing it, 'inside' and 'outside'. In this enclosure there is the hint of a tomb, of something deathly. For someone with a poetic or religious turn of mind, the figure might have a Christ-like resonance: an ancient resurrection scene in the midst of a disused and ramshackle remnant of war.

SECTION 1: VIRILIO AND THE BUNKER

Creative . . . critical . . . An image is an invitation to gaze. The image that prefaces this discussion directs attention to the bunker form. The image is accompanied by two short texts (a poem and a prose piece) which accept the invitation. Each responds by directing attention to the bunker figure in different but related ways. Each implicitly calls upon the historicality of the world in situating the bunker image. Even as they gesture toward the image, therefore, these two pieces necessarily invoke standpoints that reach beyond its 'immediate' terrain. Each rehearses traces of unseen movements of eyes embedded in practices of looking, hinting at elements that are implicit not only in the image but also in the awaiting discursive possibilities that unfurl from it. Each is a kind of confession.

The following text engages with these creative preambles, seeking to translate the terms implicit in the preceding image and texts into a more explicitly critical idiom. It does so firstly by engaging with the work of Paul

Virilio and his analysis of the bunker as an historical-social form that relates in an essential way to the fashioning of Western visuality. Principally, this relation concerns the manner in which dominant modalities of contemporary technological visuality can be read as articulations of the watchful, ancient, militaristic logic of the defensive bunker. This logic is expressed most powerfully in the fashioning of the contemporary city space. Virilio's interpretation, it is argued, is powerfully revealing but rests on a questionable metaphysical assumption about the nature of the self and its relation to technology. Virilio cleaves, in short, to a notion of the self that is transhistorical. In order to address this concern, a distinctive interpretation of Nietzsche's *The Birth of Tragedy* is offered. This reading develops Nietzsche's notions of the Apollonian, Dionysian, and Socratic to offer an account of the visual which is compatible with Virilio's understanding of hypermodern visuality as an extension of an archaic defensive watchfulness but which, at the same time, avoids its metaphysical commitments. Taken together, it is argued, the notions of the Dionysian, Apollonian, and Socratic can function as revealing elaborations of the archaic structure underlying the articulation of both modern visuality and its intensified hypermodernized variant. Principle amongst these in relation to contemporary culture is Socratic visuality, which initiates the notion of 'truthful' vision as being exemplified by the exercise of a distantiated, observational, and illuminating objective reason. Needless to say, the status of this conception of reason and the visuality it connotes is questioned. This questioning extends throughout the entirety of this volume and thus forms its backbone.

Since much of the terrain of the introductory discussion has been set out via Nietzsche and he will be turned to again in due course, it is worth noting that Virilio is not a great fan of Nietzsche.[2] Even so, as has been hinted already, they share much in common when it comes to elucidating the cultural conditions which constitute the historical 'a priori' of Western visuality. Virilio adopts a version of the approach already intimated of thinking through the inheritance of modernity in terms of its complicity with the primitive violence of the ancient social bond which returns to haunt it. In this regard, as Arthur Kroker has commented, 'Virilio *is* the truth of Nietzsche's prophecy'. Virilio's approach, moreover, allows in a powerful and productive way for practices of seeing to be brought to the fore and interrogated in their modern and hypermodernizing forms.[3] Like Nietzsche's, Virilio's thought is also starkly visual in implication. In this regard, Virilio only conforms minimally to the tendency identified by Martin Jay of twentieth-century French to the denigration of vision.[4] Like many of the thinkers Jay deals with, Virilio is suspicious of contemporary articulations of visuality. What characterizes Virilio, however, is less a suspicion of sight per se than the moulding of visuality at the hands of modern techno-culture which initiates the 'furtive dawning' of a new era of 'electro-optical light'.[5] As John Armitage and Ryan Bishop note, 'Virilio's major contribution to contemporary European thought has been to demonstrate that questions of visual culture are [. . .] extremely important political questions', for they extend across all spheres of modern life.[6] Central to this, they add, is the battlefield and the technologies which spring up within its space. One can follow this comment and approach Virilio's unique insights by turning to his analysis of the defensive structure of the bunker.

Virilio holds that when considered sufficiently closely, a bunker intrudes into the vision.[7] It is like a thorn. It irritates the eye by creating a distortion and discomfiture of space. In this, the bunker seems to encapsulate the pain and anxiety which typify the shifting uncertainty and tumultuous violent change in political and social relations which marked out what has memorably been called the 'Short Twentieth Century',[8] during which time it was designed and built. The bunker, like its times, exudes an aura of the temporary.[9] This aura is a practical consequence of its structure. The bunker 'is not really founded; it floats on ground that is not a socle for its balance, but a random and moving expanse that belongs to the oceanic expanse, and extends it'.[10] The bunker lacks the stable foundation that would make for such longevity as might allow it at some point to emanate a sense of place, of locale. It is short-lived by design. Its place is no place. Even the very land it sits upon slips and moves around it. Indeed, as a visit to some bunkers readily reveals, if left long enough the structure will tend to slide over and into the landscape, tipping up and sinking beneath or rolling over the sand or rubble on which it has been put.

Ultimately, all bunkers are doomed to vanish within the fluid surface of their placements or roll away and crumble. In this they ironically fulfil at least one of the central intentions underlying them, namely, that they become invisible to prying eyes. The lack of a stable viewpoint characteristic of the modern fashioning of vision is echoed in the bunker's structural propensity for slippage. With it, the land takes on properties reminiscent of the water on which ships float: formerly solid terrain becomes as fluid in a manner reminiscent of the sea . . .

. . . The liquidity of the defensive bunker's hidden placement is also reflected in its poured concrete construction, which has a rounded finish to help reduce the chances of a catastrophic hit by enemy fire (projectiles tend to ricochet off smooth edges, whereas they penetrate more easily into sharp, angular ones). It is often the case that smoothed-over lines indicate the presence of the slow eroding power of wind and water. This erosive power comes to endow what it weathers (rock, for example) with the quality of age. The bunker's deliberately rounded lines, however, are suggestive of something unnaturally premature. It possesses qualities of being new and yet already old, manifesting in its structure a confusion in which the hint of an archaic presence lurks within the exemplary modernity of its construction. This rounded appearance likewise facilitates the unnatural and uncanny feeling which accompanies the bunker's striving to recede and disappear into its environment so as to minimize the danger of attack. As Virilio suggests, when viewed with a suitably critical sense, the structure comes to stand out precisely because its technologically designed tendencies generate something anachronistic and irritating to the eye in their ambiguous mixture of conflicting senses of the temporal alongside the peculiarity of the structure's placement in the landscape.

Virilio notices that one particular kind of gaze, namely, that of the camera lens, seems to facilitate this critical perspective extremely effectively. The bunker's structure, being 'absolutely different from the forms of ordinary constructions', is revealed most powerfully by the interposition of the camera, for it is revealed as being 'scandalous on a snapshot'.[11] Virilio speculates that this sense of the scandalous may be related to visual awareness and the shapes which allow gas and fluid to flow unhindered over hardened surfaces. Just as water flows more easily off a structure with rounded edges so the eye is inclined to roll over and off rather than look 'into' any space circumscribed by a series of smooth curves.

This channelling of flows stands out when it is presented in a photographic image because such an image exhibits propensities that it both shares with the bunker and simultaneously makes manifest within its own organization of space. A photograph can pit background and structure against one another by bringing them openly together within the flattened-out two-dimensional plane of its field. It can throw smoothed edges into angular relief, emphasizing them compositionally by exploiting the limits of its frame, making them jut out, intrusive and unnatural. With the image of the bunker, the environment is shown by being forced to conform to a fusion of technological, compositional, and organizational demands. The background that surrounds the photographed bunker structure becomes in this way open to being drawn into its glowering presence in a way that can seem to question it.

It follows from this that there is an important sense in which, in pondering the image, one never sees the bunker 'itself' at all. The photograph is never merely a 'representation' of a bunker, a passive mirroring of its emanations of light. It is an abstract cultural concatenation consisting of practices which put flows of light to work and which, in doing so, refract the cultural-technological social world from which it springs. This world, like the one conjured by the photograph, puts to work flows, just as the bunker channels water and gas across its rounded surfaces. In the photograph this happens in a manner that makes manifest the principles of organization at work in that cultural world. One encounters thereby the principle of the bunker's construction in the image and as the image: it is a structure which in extending itself outward draws the environment into its ambit, colonizing it. This ambit is also that of the photographic image, which lends itself to the bunker. The image's composition renders its background as the bunker's background. Its space is the bunker's space. Its sight is the bunker's sight. But it does so by at the same time obstructing these elements, interpolating itself between the seer and the seen as the cost of its presentation. One can only envisage the extent to which this concealment occurs if one substitutes the bunker, and its apparent normality, with something else. Consider a state of affairs that is a little incongruous in contrast to the bunker's 'normality' . . .

The photographic logic of enclosure stands out because of the apparent unnaturalness of a situation that is, in reality, far more 'natural' than that of defensive space. The environment is not naturally ordered; things wash up at random which testify to the fallibility of biological forms of organization. What is jarring here is the manifest sense of vulnerability of the unprotected body, which is inimical to the controlling mentality which leads us to surround ourselves with protective layers. At work here is something that resonates with Anders's conception of 'Promethean shame'. Such shame contrasts with Promethean pride, which encapsulates a defiant sense of autonomy in the face of existence. Prometheus brings fire to humankind and with it the gift of the ability to manipulate and fabricate. Promethean shame is the reversal of this pride. It springs from the unflattering contrast that appears when one compares the increasing perfection evident in fabricated things with the imperfect, flawed nature of the human being, who cannot be made as a direct consequence of the perfection of technique since he or she must first be born. The person who sees themselves subjected to a comparison with 'perfect devices' becomes ashamed of 'his bodily clumsiness and his corporeal imprecision'.[12] A corpse on a beach: the pride the camera engenders encloses and renders shamefully pathetic the imperfect vulnerability of a body. In becoming aware of this, in seeing this body, one is forced to see something of oneself captured by the cleanly ordered combination of frame and lens. This is a capture which, in enclosing the body, at the same time excludes it from its ordering of space on the grounds of its lack of order and perfection.

If the bunker is revealed in a photographic image as being grotesquely unnatural it is because the photograph, too, partakes of the same grotesque and unnatural manipulation of space as the bunker itself. The camera is, after all, also an enclosure, an enclosure for the purpose of organizing light: light is 'allowed' to enter through the lens, which converts it into a flow of intersecting rays capable of being situated on a flattened plane. What the photograph reveals is that the space which the bunker occupies, which is also its own, is permeated by a specific kind of territorial principle. This principle is detectible when light is drawn through the camera lens and gathered on the sensitive, flattened-out plane behind. It is a congealed or colonized space, although one can also use a term of Virilio's and call it 'saturated'.[13] This congealed, colonized, or saturated quality, which is paradigmatic of the form of territoriality in question, is at once presented (by way of the bunker-structure) and concealed (in the very presentational qualities of the image-structure which throw the bunker-structure into relief) in the two-dimensional plane of the photographic optical illusion.

One can reflect on this image-structure by considering the photograph's ability to generate its own peculiar and impossible sense of depth, which it does precisely by denying entry to the viewer even as it presents its view to them.[14] The notion of congealment implies something soaked through and homogenized: a space that is liquefied like thick mud, viscous and cloying, yet mobile and fluid as oil; glutinously organized insofar as it is almost entirely populated and devoid of gaps which have not been penetrated into. Restless eye movements jar with this organized, resistant space even as they strive to inhabit it, which is why in viewing they tend repeatedly to traverse its artificial coordination of visual elements in a manner that, if it were to be done for long enough, might come to resemble the psychotic, circular pacing-round of a wild cat in a cage.

Virilio allots the photographic image a special place within this saturated space, one which points to matters of wider social and historical significance relating to the sphere of modern perceptual and interpretative practices. Our increasing dependence on the 'Cyclops eye of the lens'[15] has the effect of undermining memory and standardizing seeing by harmonizing both in tandem with the optic's mechanical-visual possibilities. Vision in the modern era, on this conception, becomes subordinated to the instrumentally regimented domain of mechanical representations and the communications systems which underlie their production, distribution, consumption, and, ultimately, reproduction. The hegemonic modern visual regimen which emerges in the nineteenth century is revealed here in intensified form in the twenty-first. This intensification turns on the fact that the system of industrial production which dominates Western social order becomes increasingly orientated toward the creation of images and to overseeing (managing and administering) their ever more pivotal role in the social world. Crucially, space is ever more amenable to being sliced up and apportioned according to the demands of a technology which regulates social orders on a global scale. In the words of Vilém Flusser, 'The technical images currently all around us are in the process of magically restructuring our "reality" and turning it into a "global image scenario"'.[16] The ethnicity of Western social space becomes in this way global space. It is industrial modernity intensified by way of transformation into the universality of the industrially produced, managerialized image. This is the era of the regulated image-sign, of the image as a mechanically reproducible and distributable artefact of mass production governed by the shared temporality of mass-communications devices. This is the condition of hypermodernity. Hypermodernization unfurls in tandem with the development of the mass-produced image. It is the era not only of mass production and consumption but of the mass production and consumption of the technically generated image.

The initial moments of this intensification of seeing to the point of its congealment of cultural space can be traced back at least to the outset of the Short Twentieth Century. This, to use a phrase of David Wills, marks the overt appearance of 'teletechnological' conflict, the condition of war inaugurated by motorization under which combat is no longer 'concentrated in face-to-face or, as it were, hand-to-hand combat'.[17] Teletechnology extends war fronts, undermining the notion that a point exists where conflict stops and non-combatant life begins. (This will be considered further below in relation to 'total war'.) From the time of the immediate aftermath of World War I, Virilio points out, 'the perceptual field in Europe was invaded by certain signs, representations and logotypes that were to proliferate [. . .] outside of any immediate explanatory context, like beak-nosed carp in the polluted ponds they depopulate. Geometric band-images, initials, Hitler's swastika, Charlie Chaplin's silhouette, Magritte's blue bird, or the red lips of Marilyn Monroe'.[18] All of these iconic images testify to the victory of a tendency in contemporary Western culture to embrace a condition of 'message-intensification' by way of powerful 'visual and oral' techniques[19] that flow from the '*industrialization of vision*'.[20] Saturation is linked to the cultivation of a kind of commodified image whose intensity exhibits a power to overwhelm and invade everything proximal to it.[21] The image's brightness and power makes the seer blind to everything else. These images contribute to the regulating and flattening-out of the social environment over which they come to preside.

One needs to grasp the precise nature of this militarized flattening. What is encountered here relates to territorial organization and its interweaving with sensibilities and intensities of feeling. Iconic images (company logos, flags of party allegiance, stylized representations of femininity or masculinity, etc.) overwhelm. They express an organizational principle of power directed toward the manipulation of space, each condensed like a précis into an imprint of an image-form that is readily reproducible and consequently easily communicable. Light gets the same treatment as the bunker's of liquids and gases: it flows through the social domain, helping to iron out stubborn folds that persist in increasingly smoothed-out surfaces. The iconic image replicates the principle of the bunker insofar as both come to stand out in the environment like little castles, diminutive colonial constructions which replicate the claim to the possession of space as determined visual statements that it will be defended.

To be overwhelmed by an image's intensity means to respond to it only at the level of that intensity, to hold that the appearance that confronts one at that moment is everything and that 'reality' stops there—that the reality of the 'referent' is 'captured' in and as the image itself. To be immersed in the 'referent' in such a context becomes equivalent to being suffused and colonized by the organization of light that pervades the image and so integrated within its domain. One thereby accedes to its flattened, walled, congealed visual plane. In simpler terms: the person overwhelmed by an image risks becoming an adherent, a fetishist of the icon, an idolater. The believer stops with the visible image, pausing before what confronts them as authoritative and declining to explore further the dominant principles that are being exerted over sight and, therefore, the power of domination exerted *through* sight. This summarizes the essential power underlying the contemporary advertisement, which utilizes modern visual technologies of presentation to exploit archaic visual sensibilities—the advertisement being well characterized by John Berger as the 'last moribund' form of the European post-Renaissance visual aesthetic.[22] Consider a world that becomes ever 'brighter' and more intense, more 'image-conscious' (which is not the same as being conscious of the image or of the possibilities of expression in relation to sight, possibilities which necessarily include artistic reflection upon its limits). Such a world is increasingly illuminated, suffused, and saturated with forms of light of its own technological making. It is a world that is 'en-lightened', i.e. pervaded by the saturation of a form of light that is industrialized (the light of the power station). Might this be a world no less prone to the cultivation of the invisible, to its own kinds of concealment and mythology, than the most 'unscientific' of supposedly 'primitive' social orders? One can approach this question by considering the military articulation of urban space and its link to industrialized culture.

SECTION 2: URBAN BUNKER SPACE, DROMOLOGY

One can detect the principles of industrial saturation at work in the construction of hypermodern urban space and its visual possibilities. Virilio highlights the articulation of this space's emergence in the eighteenth century, its initial explosive political expression exemplified by the shocking modernity of the French Revolution.[23] Although he does not mention it, this space is also an unfolding of that century's initiation of the industrial revolution's division of labour, outlined so memorably by Adam Smith.

As is evident in Smith's well-known advocacy of the so-called free market, modern political fervour springs from the assumption of an autonomously organized free urban space where freedom and the pursuit of profit are merged into one and the same meaning. The assertion of this free space is bound up with the notion that the city is the radical source and centre from which intellectual illumination and material advancement are spread. Change is conceptualized as radiating outward from this centre via commercially articulated webs of connections toward more traditional, shadowy, backward rural areas.

A famous passage in *The Wealth of Nations*, which elaborates how the varieties of manufacture concealed within a single individual garment make it a microcosmic illustration of the advantages of the division of labour under the urbanizing aegis of industrialization, demonstrates Smith's adherence to the above conception of urban cultural priority and superiority.[24] Here, Smith's presupposition about the primacy of urban-industrial production as a positive driver of change is well revealed by the language he resorts to when he expresses his view of its relation to and effect upon the rural human source from which the labouring classes are drawn. Even the apparently 'coarse and rough' 'woollen coat [. . .] which covers the day labourer', Smith comments, reveals a level of metropolitan refinement that belies the identity of its crude, uncivilized rural wearer since the coat is 'the produce of the joint labour of a great multitude of workmen' whose actions have been organized and coordinated by the superior urban powers of organized capital.[25]

Smith depicts the labourer as a creature adorned with a mere 'cover'. In so doing, he conjures an image of the creature concealed beneath the coat as being akin to a primitive animal. The clothing which wraps itself around the labourer reveals the essence of the being it conceals. Clothing here is a symbolic and starkly visible manifestation of the culturally superior powers of urban organization which permeate and govern the day labourer's rural mental world. Smith's account enacts in textual form the enclosing and encoding of the labourer's body within the realm of the visible in terms of the products that he or she is enlisted to produce within the capitalist process. Each of the coat's woven threads charts the labourer's inexorable incorporation into an environment in which he or she is deemed fit to be seen and worthy of comment only in virtue of having conformed to the saturated homogenizing condition of organized production and consumption which inexorably encloses them.

Such a view, so influential that it has become a kind of rote wisdom, presupposes more than it has warrant for. Above all, it supposes that the cosmopolitan commercial space of capitalist production can be treated as an independent social milieu. The city, however, is amenable to being regarded as less of an organizing centre and more of an instrumental conduit than one might think. 'A modern city', as John Berger puts it, 'is not only a place, it is also in itself [. . .] a series of images, a circuit of messages'.[26] Cities, in other words, emerge as spaces because of the information that flows through them; they are not inert, neutral geographical spaces constantly waiting to be 'populated' by varieties of images (visual, material, and mental). The city space itself, rather, is a product of the flows. This is evident in the city's origins in practices that continue to motivate it and which, Virilio notes, spring from its location as a convenient stop-over, a geometrical point, a site defined by a criss-crossing of channels of communication (a necessity for the transportation of goods and people) which in converging, flowing over, flowing through, and entwining with one another serve ultimately to establish it and make it stand out as a distinctive space.

From the outset, elements of modernity are implicit with the city, for it has always been girded by the temporary. It has always been a place of meeting and moving on characterized by the presence of 'river, road, coastline, [and] railway'.[27] A modern city street illustrates this well enough. It is no meandering wood-path that, should we choose to follow it, takes us where it will, even to the dead ends shrouded by brambles beloved of Heidegger.[28] As Friedrich Kittler notes, 'Founding a capital city today simply means that at freeway interchanges and railway stations, in timetables and computer networks, a new "hub" emerges, which centralizes the flow of energy and information'.[29] It is easy to fail to see the urban wood for the trees and ignore the primacy of flows in favour of the social agglomerations which accrue in virtue of (and around) them. Streets and commercial areas can take on the appearance of 'centres', but like sewers they are also hubs and conduits for containing, passing, and directing materials of various kinds on their way to somewhere else. What typifies them as much as cosmopolitan dwelling, and what both precedes that dwelling and is sustained within it, is the practice of the management of varieties of flow. As with streets and sewers, most of what enters a city generally does so in order ultimately to pass back out of it. Such passing through constitutes the vista of the cityscape in accord with the demand that what flows in and out is always amenable to observation and regulation even if, as with contemporary Western sewers, this is concealed from everyday view. A city's flows are what make it what it is, and they are, moreover, matters of concern that relate more essentially to security than the cosmopolitan attitude of openness which swirls along supported by their various undercurrents.

It is not much of an effort of imagination to notice how flows of language and other actions which get associated with expressions of cosmopolitan freedom can be mapped easily enough onto conventions with more authoritarian associations. Given its propensity to a concern with security and defence, 'freedom of speech' is always an object of military awareness, which is one reason why the free flow of speech necessarily gets subjected to the need for regulation in even the most 'open' ('liberal') of metropolitan milieus. From a military standpoint, what matters is the regulation of movement. It is this which, for Virilio, constitutes the founding of a metropolis: 'The city is but a stopover, a point on the synoptic path of a trajectory, the ancient military glacis, ridge road, frontier or riverbank' originating 'where the spectator's glance and the vehicle's speed of displacement were instrumentally linked'.[30] Not the garrulous expression and openness of the *agora* (the market place—an image conventionally associated with the lively social realm of that paradigm of city life the ancient Greek *polis*) but silent observation and covert, efficient communication establish the city, drawing out its spatial potential as an extrapolation of its observational utility.[31]

This condition is effectively summed up in relation to contemporary hypermodern social order by Virilio's conception of the 'dromosphere'. Dromology (from the Greek *dromos*, meaning 'race') is the discipline that Virilio proposes for the formal study of this phenomenon. Dromocratic social orders are social domains governed by the demand for ever-increasing speed, of which the efficient management of flows is a prime example. They make manifest the occidental faith in its own technological pursuit of superiority over nature and can be summed up by the maxim 'Speed is the hope of the West'.[32] The phenomenon of acceleration is illustrated well enough by considering the motorization of the horse-drawn carriage, which revolutionized social and political spaces by reconfiguring spatiotemporal relations between locales in terms of accelerated flows of production and consumption. Further acceleration in late modernity heralds the advent of communications systems which motorize cultural space by regulating it in terms of ever faster synchronized flows of information.[33] That the desire for speed has an undeniable power is evident from its ability to shape social and natural environments in decisive ways. Speed is convenience. Speed is control. Speed means being there first, seeing first. Speed is domination. Speed is likewise, in modern times, the central driver of global commerce and war. It also threatens the undoing of politics and critical reflection.[34]

Understood in these terms, urban environments lend themselves to being conceptualized as unfolding within the privileging of a specifically dromocratic historical articulation of the visual which conjoins habitually cultivated movements of the eye with the accelerated and coordinated movements of traffic.[35] This mode of visualization turns on the instrumental regulation of the ancient defensive relation between observer and seen and harmonizes them according to the requirement that movement is always subordinated to the demand that it be amenable to being rapidly surveyed and efficiently regulated with a view to eliminating threats.[36]

The city is the archetype of the bunker. The land around it rises up, the sky above bears down, precisely in virtue of its defensive elaboration of visual space. That the bunker itself tends to come into view as something positioned away from urban sprawl, floating on its precarious fluid base on distant sea-fronts or cliff-edges, is in this regard hardly relevant. The enclosure is city in miniature. The bunker's testimony to the threat of war and the habits of visualization in response to the demand for rapid conceptualization and response are never far away from the city, as is suggested by the dominance of military elements in the most apparently 'civilian' of liberal-democratic governmental forms.[37] The bunker's realm, replete with its striving for concealment, is merely a glaring extension of the saturated logic at work in a still-active ancient metropolitan tendency to pursue practices of observation. It is an exemplar of the urban defensive inclination first to wind space in on itself (the protective wall, the chain of observation, the chain of command, rules governing rites of passage 'in' and 'out', etc.) and then extend that wound-in space in the form of an effusion of structures scattered through the wider landscape, constituting the broader environment as a dotted constellation of interlinked points, each of which gets situated in relation to the urban totality that seeks to proclaim sovereignty over it. The bunker's ubiquity[38] in the Short Twentieth Century can in this way be taken as a proclamation of the modern, limitless extension of much older practices which inhere within the urban condition.

This ubiquity is also a precursor of the bunker's subsequent return 'home' in the liquid modernization which characterizes the post-war era with its reappearance in contemporary city space. The bunker rationale is made manifest in a plethora of contexts—the protective enclosures which spring up in North America in the form of gated communities, or the concrete corridor opened to allow movement between Israel and Gaza in 1999.[39] It is also present in the heart of the city itself, where bunker-logic rises up to the vertical in 'super-gated', luxury apartment towers[40] or within the structures of buildings which combine entrances for segregated luxury apartments with 'poor doors' for tenants of social housing who live in the same building. Defence and the accompanying demand for a space which can be subject to constant surveillance, it seems, no longer needs to be catered for at the limits of national space. The promise of danger has visited itself on the city as something that pervades the possibilities of its spatiality by entwining itself within its heart, dividing it internally by way of forms which guide and divide lines of vision, revealing and concealing in equal measure.

That this tendency comes increasingly to the fore is testified to by the development of hypermodern surveillance technologies which make the ability to regulate flows ever more efficient. Cameras mounted on street corners, in shops, precincts, and on traffic lights are machines whose undiscriminating gazes observe and record the daily passing of faces and number plates alongside events of a less-routine nature and do so all in the name of 'security'. In them, the visual homogenizing of space in the contemporary urban environment is caught in the act. These features of the seen environment, which ostensibly get established in order to observe movement within it, meld with the very milieu they are meant to observe. They become endowed with the power to shape that milieu. Here one can detect at work the cultivation of powerful visual techniques which map out and so come to determine the meaning of 'lived space' in terms of principles shared in common with the regulation of flows which characterizes the grotesqueness of the bunker or the distantiated gaze of the flattened-out plane of the photograph.[41] This commonality relates to the establishment of principles relating the visible environment to delocalized coordinates, to points that relate to abstract grid references capable of simplifying and regulating flows, overseen by imaging systems.[42]

SECTION 3: TERROR, URBAN SPACE, HYPERMODERN VISUALITY

From the moment when city space is opened out, a defensive slope of ground rising up around the site of a river or other feature, its being formed in relation to the need to ward off external incursions reveals the threat of invasion as intrinsic to its existence. In this sense, the city's inside *is* its outside. Its inner logic depends upon its prophylactic relation to externality. Put another way, its peace is also war. The city's addiction to visibility is, in consequence, bound up with the constant presence of an invisible threat: if the intensified, saturated, hypermodern urban realm ends up being a place which never sleeps, it is in no small part because this is an extension of the ancient tendency to the cultivation of permanent wakefulness which is a prerequisite of the vigilant visualization proper to defence.

The hypermodern metropolis's primary domain is what is seen; it is always incessantly translating its elements and features into visual terms. Existence is in this way inextricably conjoined with a ceaseless exposure which is energized through the channelling of energy into the equipment waiting to display it—the illuminated screen, the pulsing neon light bar. The city's topography is pure illumination. It is through this, too, that the contemporary metropolis exposes itself as having a decisive vulnerability. The energy that lights the illuminated sign or directs watchful street or shopping centre lenses, like the waiting mobile phone or the cash machine camera, can be turned back on the defensive loop of sensibility (fear) that motivates surveillance. This happens most apparently in the case of images of violence: because of the potentiality of technologies of observation, images of violence inspiring fear become latent in the city's social milieu. They are always just waiting to be captured: their potential is prefigured and even invited by recording machinery. The blast of a bomb or the attack with a car or aircraft which has been turned into a weapon: these acts in simply being committed are now amenable to being automatically preserved and repeated endlessly. The violence of such acts thus gets extended and compounded by way of their ability to usurp and dominate the visually fixated urban domain, since in being preserved they automatically await collection and assembly for dissemination by studios and newsfeeds onto waiting television and computer screens.

The word 'screen' as it has just been employed, like Walter Benjamin's '*Bild*',[43] does not turn on a pre-existing distinction between the mental, visual, and material realms. As has been noted with regard to Virilio's own use of the word 'screen', it needs foremost to be understood as a metaphor[44] for the cultural condition which encompasses all of these dimensions. The realm of the metaphorical screen is the realm in which hypermodern metropolitan space and time in its visual, mental, and material registers is articulated. Cultural life is conducted in virtue of this articulation, which saturates and so energises the social relations that become bound up with it. This condition of saturation is intimately related to the contemporary form of terror. As an environment so saturated by light that it lives up to the cliché and 'never sleeps',[45] urban space is prey to the lurid visual publicity that now necessarily and instantaneously accompanies all outrage and tragedy when they are luminously 'represented' on the screen of the electronic device.[46] Gripped by the screen's cultivation of this endless circulation, the watchful gaze becomes unable to close down; like a lidless eye, it exists in a constant state of preparedness for shock.

Virilio theorises the emergence of this contemporary visual space in terms of an historical shift from a geopolitical to a metro-political paradigm.[47] Where in geopolitics geographical factors are the fundamental drivers determining conflicts between states (i.e. disputes over conditions relating to climate, sea-access, fertility of land, and the like), the metro-political era witnesses a historical return to the situation exemplified by the besieged ancient Greek *polis*, portrayed memorably enough by Thucydides in his account of the Peloponnesian War of 431 to 404 BC.[48] This is the condition of total war, which Carl von Clausewitz (writing sometime between 1816 and 1830) likewise analyses as an archaic return to the ancient Greek paradigm[49] in the Europe of the Napoleonic period in which industrialized and proto-industrializing social forms are pitted against one another.[50]

Understood in Clausewitz's terms, total war is a conflict in which not merely entire populations but also the entire gamut of social forms (science and technology, art and sensibility, economic production and trade) are co-opted into a gargantuan destructive struggle.[51] It is through this co-opting that the Napoleonic period initiates a decisive shift toward fusing the ancient total war scenario with industrial modernity. Through the nineteenth century this shift is intensified to the point where, in the twentieth, it is signalled by the bunker structure's appearance in city space as a prosthetic aid to civilian existence. Initially, this appearance takes the form of an additional skin to protect vulnerable bodies from the mass assault of bombs in air raids such as the British Anderson Shelter of World War II, designed to be assembled in the back garden of a person's house.[52]

The arrival of the bunker signals the moment when the battlefield comes to invade metropolitan space, collapsing the boundaries which separate 'inside' from 'outside'. Total war, which was assumed to be antithetical to the classical, mercantile, metropolitan ethos (and so best kept at a distance from 'civilized' everyday civic life with its pursuit of trade), thereby reveals itself as a powerful determinant of city space. With hypermodernization this condition of total war is intensified and naturalized: the battlefield engulfs the city, increasingly taking on the appearance of an everyday, 'peacetime' aspect of existence. What is assumed to be out of view, *outside* the city—the condition of war which remains beyond city frontiers so long as they are not overrun by siege—now appears scattered about within it in the form of discrete, seed-like bunker structures. The gated luxury building or the protective glass walls and steel spikes that now surround the Eiffel Tower in Paris demonstrate well enough the increased policing of urban space by various forms of blockade. Numbered amongst the latter are also additional forms of 'street furniture'—barriers and bollards of all kinds, including strategically placed, sizeable 'plant holders' and 'sculptures' whose aesthetic appeal is designed to reduce public anxiety by titillating the eye so to divert attention from their protective function.[53]

Blockades are an admission. They testify to the presence of a potentially uncontrollable destructive threat coursing and accelerating through the urban region. The threat of terror is met with strategies which reflect this. The blockade is a brake: it seeks to hinder the acceleration which precedes an attack by reasserting the distance separating project and target which is the hallmark of security. Another option is to counter one kind of acceleration with another by deploying mobile barriers which also slow things down by patrolling areas and regulating movements. The hypermodernized urban environment is now in consequence also faced with the prospect of being populated by 'security robots', mobile bunkers in the shape of autonomous machines endowed with 'ground level eyes and ears' that, one manufacturer boasts, 'enhance situational awareness like never before'.[54] Work done under the auspices of such bodies as Efus (European Forum for Urban Security) and media4sec[55] also reveals the extent to which this barrier function is already closely entwined with social media. The burgeoning universality of the blockade testifies to an enveloping state of siege. Under the threat of terrorist activity, the hypermodern city becomes identified as a terrain of conflict, a warzone; its inhabitants, whom its rulers are bound to protect in order to maintain their own legitimacy, are increasingly enemy combatants deemed 'legitimate' targets of violence. In short, the citizen—like all combatants in conflict—becomes a passive object of political struggle rather than an active subject underlying political action as liberal theory would have him or her be. (To comprehend the precise nature of this crisis, one should recall that the Hobbesian logic of the nation state is rooted in the contract which holds legitimate power to exist only in virtue of its ability to protect those subject to it from external harm.)

One can return at this juncture to the image with which this discussion began. Viewed in the modesty of two dimensions, photographically frozen in an anonymous and hard-to-pinpoint location distant from the iconic landmarks which come to define cityscapes,[56] the bunker structure, a miniature fortress, a shelter and prosthetic skin, reflects the ongoing saturation of space that emanates from the energized and light-filled hypermodern city's regulation of flows. The deserted bunker when it is viewed in the image encapsulates a frozen and deadened cityscape. It reminds us that the city always stands in need of defence and this need invades and saturates it with an observational watchfulness. The bunker stands, at the same time, as a symbolic reminder of the dangers this saturation poses since these protective visual flows can always be turned back on the city itself (no conduit is ever a one-way street), necessitating the appearance of the bunker in the city's midst. The reappearance of the bunker form *within* the bunker: this is a rendering visible as something particular of the city's invisible, universal, structural imperative.

A 'snapshot' (a word with its own connotations of instant violence) shows this saturated condition at work in the bunker structure by presenting its idea in paradoxically flattened relief that renders it disturbing. Consider, Virilio comments, one common feature of the North American street that too many people remain indifferent toward: the sight of a gun-shop display window. Revulsion for the violence implicit in such commerce is displaced easily enough to apparently distant forms, not least a defensive building like a bunker, which expresses 'the essence of arms'[57] but in such a way that it is kept at a safe, emotional distance from the everyday. To the extent that the site of the bunker is not perceived as being 'homely', violence is permitted to be seen and acknowledged without the feelings of discomfort that acknowledgement of its presence at the hearth must provoke. This is discomfort in the face of the saturated, congealed space of watchful provision. Of this, one can indeed say, 'The void no longer exists, everything can move, arrive, or go; the earth has lost its materialness, and space its emptiness'.[58] Regarded in this way, the structural-sepulchral element revealed in the photographic image's complicit showing of it serves as an allegory of the ancient historical inheritance of violence which erupts once more in the light-saturated milieu of hypermodernity.

It is worth noting that Virilio's account arrives at its conception of hypermodernized, technologized culture as autochthonous violence by constructing an opposition. On the one hand, there is a domain of human identity, sociality, and visuality which is untouched by technology. On the other hand, there is the historical-technical-militarized power of technical-scientific forms which impinges upon and distorts this identity. Thus, he claims that 'dromocratic intelligence is not exercised against a more or less determined military adversary, but as a permanent assault on the world, and through it, on human nature'.[59] It has in this regard rightly been said that Virilio 'sees the self as a kind of virtually targeted ground-zero' whose capacity for ethical relationships is eroded by the dromocratic, hypermodernizing, technological tendency.[60] Human nature, on this conception, is the victim of a manipulative thought and constitutes both that manipulation's object and point of resistance.[61] This, along with Virilio's avowed religious commitments, has led some to characterize him as a kind of anarchic metaphysical liberal[62] who, in his determination to invoke the 'pathos of origin',[63] reproduces 'the ideal of the unified humanist subject as a theoretical norm'.[64] Such metaphysical commitments have in turn led to the criticism that Virilio has a tendency to underplay or even avoid the political implications of his approach, not least in relation to the link between contemporary technology and capitalism.[65]

If one is not sympathetic to Virilio's metaphysical commitments, his position looks somewhat compromised. Even so, there is at least one reason for affirming his standpoint, namely, its productive consequences: it reveals the value that an understanding of religious practice has for grasping the nature of contemporary technical cultural forms. Virilio makes the point that 'without religious culture, it doesn't matter which, one will never be able to understand technological culture' since the 'new technologies' which shape the contemporary world 'bring into effect the three traditional characteristics of the divine: ubiquity, instantaneity and immediacy'.[66] The tendency within technology to cultivate an all-encompassing, rapidly synchronized world contains a concealed religious ambition. Technical-scientific social forms inherit a desire for the transcendence and domination of nature which has old and enduring theological roots. The important contention here is that what is at stake in our relation to religion is a matter of cultural literacy, of the kind of literacy concerning the past without which one cannot effectively formulate a critical relation to the present rather than simply this or that metaphysical commitment. The notion that one simply 'leaves behind' ancient habits and prejudices by embracing a modern, secular outlook is naive: the more non-secular historical elements are suppressed within secular culture, the more powerful these elements become. In consequence of this, scientific culture (which is not entirely the same as science) exhibits an un-self-critical tendency to foster secular religion, as is evidenced by the degree to which it conceals within itself the metaphysical-imperialist ambition of attaining a God-like status of control over the environment and, ultimately, humankind. An unhistorical approach to science allows this religious ethos to go unchallenged on secular grounds.

One can approach this issue in slightly different terms by appreciating a point made by Bernard Stiegler. Stiegler argues that insofar as there is a human world and a human 'nature', this is so only because humankind engages with both its environment and itself in virtue of an originary technical relation to it.[67] There is, on such a conception, no human essence that exists independently of tool use in all its forms (technicity) and all that is associated with it. Thus the term 'human' does not signify a naturally given entity but a form of life which emerges in virtue of a cultural-technical formation. Technicity is, on such a conception, as ancient as the human itself—not something which distorts the purity of the human form but a grounding historical condition in virtue of which the problem of the human itself (of 'human nature') arises. Virilio's view, although fecund, leads to a position that is only persuasive if one is amenable to metaphysical commitments about human nature and historical claims about the relation between the human and the technical which Stiegler's position renders questionable. Human dignity, when it is located in a sphere that lies outside human historicality, rather than being grasped as an historical achievement, displays a tendency to being chimerical and impossible to articulate unless one relies on faith.[68] Virilio's articulation of the sources of hypermodern saturation and the visualization that epitomizes it consequently stands in need of elaboration. One needs, in short, to consider the sources of the self and the domains of the visual that emerge alongside it in a non-metaphysical way. In order to do so, one can return to Nietzsche, although this time in a slightly different register.

Nietzsche's account of the creditor-debtor relationship has already been discussed above. This affords one point of entry into thinking through the origins of the human in non-metaphysical terms (that is, in terms of inheritances of practices) which guide the eye. Another complementary point of entry for thinking through the emergence of Western technical rationality, the cultural fashioning of human nature and its fundamental relation to the domain of the visible (and in turn its role in hypermodernization) is presented in what Virilio regards as Nietzsche's 'only book',[69] which was also his first, *The Birth of Tragedy*.

SECTION 4: FROM ARCHAIC VISUALITY TO TECHNOLOGICAL RATIONALITY—APOLLONIAN, DIONYSIAN, SOCRATIC

The Birth of Tragedy's enabling supposition is seemingly nostalgic. It contends that ancient Greek tragic drama (epitomized by the works of Aeschylus and Sophocles) is a highpoint of cultural achievement in relation to which all that is modern can be judged and found wanting. It will turn out that Nietzsche's argument is that the flawed modern milieu has abidingly archaic roots, but for the moment it is worth charting the text's initial laying out of its terms. The text's terms of engagement are apparently metaphysical. Nietzsche argues that ancient Greek culture forged a productive relationship with mythical understanding through art, which rather than morality is the genuine metaphysical activity of human beings.[70] In tragedy, he argues, one discovers a unique blending of opposed mythically articulated forces. As Friedrich Kittler notes, the reason for turning to the mythical register resides in the fact that Nietzsche is engaging with something that cannot be recuperated by straightforward analytical language: 'To express what is at issue, mythical names [. . .] are required, as well as a physiological parallel'.[71] Myth, in other words, affords a point of entry into a problematic which centres on the body and its functions (cultural and biological—assuming that these two strands can be easily disentangled from one another). Myth conjoins with the body: Nietzsche, in order to initiate his discussion, calls upon aesthetic principles 'which burst forth from nature herself, *without the mediation of the human artist*'.[72] First, there is the illusory and imagistic Apollonian principle of individuation. This is law-like, conventional, associated with sculpture and architecture. The realm of the *polis* is given order by this principle's social function as a normative regulator of daily life. Second, there is the Dionysian elemental force of natural excess. This denotes an embodied condition which is perverse, revolutionary, and anarchic. It is associated with bodily drives and excess, festive revelry, and the overturning of established social norms.[73]

Attic tragedy turns on the creative tension which erupts when these contraries are brought together. In the single discursive space of tragic drama they are forced into an uneasy coexistence. The traumatic effects of their meeting and simultaneous repulsion engender an art able to confront and convey the violent conditions of existence. Since the tensions presented in tragic art parallel the tensions and strife at work in everyday life, tragedy conveys the violence underlying life, but it does so in a manner that is bearable. The bitter pill of being here is sweetened by the piquancy of the tragic, which throws an Apollonian veil of poetic language over the violent and chaotic reality of Dionysian events. In the case of Sophocles' *Oedipus Tyrannus*, for instance, these events concern the perversity of the hero's destiny. Oedipus will unwittingly be compelled into the murder of his father, marrying his mother, and siring his own siblings (whose uncle he also is), but this fate he can do nothing to avoid in spite of his repeated attempts to do so. In presenting this appalling unravelling of events as art, the drama endows even what is terrible, unsettling, and unnatural with the aura of a redemptive and seemingly timeless wisdom generated by aesthetic feeling. The suffering to which all humanity is heir is revealed in all its absurdity and excess by being contained within a strictly delimited and controlled aesthetic form. This suffering is made manifest as the silences in the play safely open up vistas which allow for reflection on destructive Dionysian abysses of doom by containing them between bridging structures of elegant outbursts of Apollonian poetic expression. Here, for Nietzsche, is a mythic celebration of life in all its pain by poetic means that affirms and so invests cultural identity with a sense of its own future even in the face of torturous absurdity. This is the great achievement of Greek civilization: it learned to live with itself and accept the contradictory poly-dimensionality of its experience.[74] The drama of *Oedipus* and its complex relation to the theme of vision is engaged with in detail in part 2 chapter 4 of this volume. For current purposes, it is worth pondering in more depth some of the implications of Nietzsche's argument in *The Birth of Tragedy*. There is an undeveloped line of thought implicit in Nietzsche's interpretation which can be taken further.

What is striking about *The Birth of Tragedy* is the central role the visual plays in it in relation to practices which give rise to the formation of social space and self-disclosure. Apollonian sensibility concerns the domain of the visual, which emerges as a space of appearance with the power to organize concrete bodies. Apollonian space is organized and regulated space: it generates the veil of civic peace (which leads to the rule of law) that fends off the chaotic external environment. This is revealed readily enough when one ponders the way in which the Apollonian is held by Nietzsche to epitomize the principle of individuation. Individuation establishes the sphere of social life, the *polis*. It does so by establishing it as a domain in which self-reflexive persons emerge (for personhood is a social construction) both surrounded by and as observers of rules (laws, customs, conventions, and the like). The individuated self is an Apollonian rule-follower, a being of technical propensities insofar as he or she is the outcome of an inculcating of formal rules and regulations determining appropriateness within shared social space. One might say that the Apollonian self is what Nietzsche later, in *On the Genealogy of Morality*, calls the 'sovereign individual'.[75] Such a being is the product of the most primitive social relationship there is, that between creditor and debtor, a calculator who is the follower of rules who gains a sense of autonomy and individuality from conformity and the sense of power it endows them with over themselves and others.

The Apollonian self is, ontologically speaking, an illusion since it springs from bodies following conventions, calculating, reckoning, coming together in the shared enclosure of social space. This self cannot exist independently of the social domain: he or she would vanish with society's vanishing. (Such a being should not be confused with the biological individual, the existence of which is not incompatible with lacking all signs of 'individuality'.) There is, if one develops the implications of this approach, no need for attributing a metaphysical underpinning to the self in the shape of any a-historical soul-substance that precedes the conditions of its formation. One can, in other words, account for the fashioning of subjectivity in terms that do not rely, as Virilio does, on metaphysical presuppositions about the nature of the human as pre-existing the technical realm. Indeed, the technical and the self emerge in tandem. The individual comes to existence through social life[76] as a counter, calculator, and observer and does so precisely in virtue of the fact that he or she has been subjected to the ruling power of techniques of observation.

One can elaborate this point relatively straightforwardly. To be a social being (a person) is to be one who conforms to the laws which constitute the social bond. This entails being seen to do so and to reckon accordingly (to live by and apply the law both in relation to others and oneself). To be a 'citizen', which is the political expression of this condition, is to be a public being, a creature encapsulated by 'publicity', that is, by an awareness of public visibility in virtue of being both observer and observed. The Apollonian gaze thus comes to radiate inward and outward at one and the same time, forming a kind of circuit of the visible. This visibility is celebrated in the Apollonian aesthetic tendency to emphasize the individuated human form. Understood in these terms, Apollonian subjectification is revealed as the founding articulation of cosmopolitan organization. As a being who is seen, one becomes also, at the same time, a being who sees and thereby *oversees* the public realm that all partake in. Social power, one might say, appears in tandem with a calculative optics which, through its affirmation of the space of the city (the *polis*) arising out of practices orientated around public observation and the charting of environed space, constitutes publicly visible space as the space of the self.

Apollonian social existence can feel like a prison. In the ancient Dionysian festivals, freedom and escape are intimated as the 'hostile barriers that necessity, caprice, or "imprudent convention" have fixed between persons' and are cast aside.[77] To the extent that it is held to be musical, auditory, tactile, rhythmic, and excessive in its pain and contradiction,[78] the violent and incalculable Dionysian strife which breaks through these public norms seems to emerge in Nietzsche's text as the contrary of the Apollonian visual ethos. In the frenzy of Dionysian excess, a 'higher commonality' is revealed than that of the individuated life of the *polis*. This commonality lies in the persistence of a link which conjoins the embodied existence of humankind with the natural environment from which it originally sprung but from which it has alienated itself by entering the technically regulated, observational world of civilization. Through Dionysus the uncivilized body shudders with an uncontrollable, invisible life-force that threatens to explode the sociality of public existence and civic peace.

The Birth of Tragedy does not, however, turn simply on the opposition between the 'visual' and the 'acoustic-gestural'.[79] Dionysian sensibility opens up cracks in civilized social space. It does so, moreover, not by invoking the immediately tactile realm of the body but by asserting another viewpoint in the shape of an alternative modality of seeing and being seen that emerges within Apollonian social space. Nietzsche's account of the role of the chorus in Greek tragedy points to this modality. Even a casual reader of Sophocles' *Oedipus Tyrannus* can hardly fail to notice that the chorus comments on the unfolding events of the play. For this reason, the chorus might be concluded to be a kind of idealized audience of detached observers. Nietzsche argues it is not. A public audience understood as a body of disconnected and passive onlookers 'was unknown to the Greeks', as is revealed by the design of their theatres.[80] Rather, the chorus is a 'Dionysian mass of spectators' presented in mythical form as satyrs. Satyrs are 'restored geniuses of nature', beings who symbolize the unregulated realm of bodily desires that lies outside the control of the *polis* but exists within its domain in the embodied conditions which underlie all socially individuated existence. The satyr chorus is not primarily tactile, however, but expresses above all a mode of 'vision'. The viewer of the play is drawn into the sphere of this vision so that he or she becomes a member of the satyr chorus and a participant in and witness to the unfolding events of the play. Thus, in spite of its association with music, the Dionysian generates a sense of space that is, like the Apollonian, fundamentally visually orientated.[81] However, where Apollonian sensibility enacts a social world in which identities are constituted from a domain in which seeing is organized around the regulated, law-like behaviour of individuating public space, Dionysian sensibility enacts a vision of excess that calls upon this regulation in order to overrun and suspend it.[82]

The vision of the chorus enacts a mythical over-spilling through which the Dionysian crowds out the Apollonian realm of measurable and lawful everyday civic life by engulfing it within a swarming invasion of poetically imagined, luminous figures. The distinctive energy of these figures springs from a mode of sight which draws on a mythical imaginary that invokes an ecstatic absorption capable of suspending the primacy of individuated social identity. This visuality initiates a seeing and thinking different from the norm which emerges precisely from its engagement with the norm. In it one is no longer 'oneself', for the notion of the individual is shown to be illusory as it recedes and dissolves into the choric unity of ecstatic expression. The relation to language illustrates this condition well enough. In Dionysian ecstasy, language becomes invaded by a force of mimetic sight: the 'genuine poet' who unleashes Dionysian powers does not regard metaphor as 'a rhetorical figure but [as] a vicarious image that he actually beholds in place of a concept'.[83] Poetic thought, in other words, is differentiated from individuated, regulated consciousness by the infectious quality of its disruptive powers of visualization, above all, by its ability to flood the regulated world of public observance with an overflowing of unregulated image-thoughts which defy the world of sense as stipulated by the conventions of everyday law and custom. In this defiance, the Dionysian takes on an excessive quality containing intimations of terror and threats of violation, for through its ecstasy of image the interiorized condition of law-like behaviour which articulates the space of the *polis* is de-encoded and so made vulnerable to the forces it was formed to ward off.

The realm of the seen considered in these Dionysian terms is no ordered representation of the environment. It is creatively presentational and bodily rather than observationally and abstractly representational. It is the imagistic domain as felt (as haptical) rather than representationally thought (as optical). In terms of tragic drama, this creativity is made manifest in the Dionysian's power to transform the seer (the viewer of the play) into one capable of beholding the god on the stage. To see in this rapturous religious condition becomes to see with Dionysian eyes, that is, from a standpoint 'beyond' everyday rule-following.

Social order is a concatenation of primitive urges brought to heel by the power of convention. Dionysus stands for the artistic condition through which these urges are once again allowed expression: the worshipper of Dionysus becomes animated in a manner that transcends the observational power of that order. In the Dionysian state the visual merges with the tactile rather than standing as observationally distanced from the tactile as it must do in individuated Apollonian experience.[84] In such a condition, seeing itself becomes integrated into bodily experience as a form of touch capable of connecting all the visualized points it beholds to one another through sensibility and so dissolving them in feelings of ecstasy. Instead of discretely visualized points 'in space', what is seen is beheld as a more generally perceived (i.e. felt) constellation. In this, it runs counter to the disconnected optical separation which is characteristic of the self-regarding, rule-following Apollonian realm of metropolitan existence.

Nietzsche has, with this, offered not so much an account of the metaphysical ground of classical aesthetics (which is what the text of *The Birth of Tragedy* promises its readers) as a post-metaphysical account of the cultural construction of the Western self in virtue of the historical institution of dominant structures of practices. There is no need on this account, as there is with Virilio, to appeal to a universal conception of human nature to ground the social realm and the subjectivity which dwells within it. The self emerges, rather, as a non-metaphysical Apollonian policing and directing of resistant Dionysian bodily tendencies. Personhood is, to this extent, an unwitting cultural achievement which does not need to be referred to the authority of any higher trans-historical category in order to be what it is. Such a being can suffer (in virtue of his or her Dionysian attributes) and be both a victim and perpetrator of wrongs (in virtue of his or her Apollonian attribute of being subject to the demands of custom and law).

The organized and measured life of the *polis* is not Dionysian ecstatic rapture's deadly enemy, however. In order to flourish, Apollonian civic life depends on Dionysian excess: the existence of the socially regulated world of the *polis* presupposes the bodily forces which it regulates and oversees, and excludes and represses, in order to be possible (social life is the organization and cultivation of bodies, after all). Equally, without Apollonian constraints the Dionysian would be no more than untamed animality, a shattering of culture rather than its powerful, passional source of expression. The relation between the Dionysian and Apollonian is one of contestation, but not with the aim of mutual annihilation. The enemy of the Dionysian is, rather, the form of sensibility that Nietzsche associates with the inheritance of modern times. It is a sensibility which undermines the unstable but productive balance of Apollonian and Dionysian elements.

To be modern, Nietzsche notes, is to embrace the ethos of a passive observer rather than an involved one. It is, one might say, to retreat from the walls and battlements of the Apollonian enclosure of the *polis* in order to police them not by moving through them concretely but by attempting to detach oneself from their environment by way of a distantiated and calculating gaze. The paradigmatic modern self, Nietzsche comments, reflects this. The modern person is unable to lose him- or herself in ecstasies in the manner that Dionysian experience demands. This is because we moderns have become uncreative collectors and cataloguers of cultural experiences who engage with them by holding them at a distance, pondering and counting rather concretely participating in them.[85] Like the tendencies associated with Apollonian and Dionysian sensibility, this modern condition of detachment also has ancient origins. It emerges, Nietzsche argues, in the ancient Greek world at the moment when the contested cultural space marked out by the tragic genre is compromised by the introduction of another form of visualization. This is the cultural emanation which goes under the name of 'Socratism' and which accompanies the invention of what some might consider philosophy proper. With Socrates, vision becomes conjoined with the idea of conceptual transparency, i.e. with the idea that to make things visible means to know them, to see through into their innermost workings by way of conceptual tools which order them according to the dictates of dialectical thought and the rule of number ('logic').

Socratism opposes the ecstatic vision of the Dionysian chorus with the standpoint of a single 'rational' and judgemental spectator. One ought not to confuse the visuality associated with Apollo—the god of sunlight and a metaphorical articulation of the principle which organizes the *polis* as a space of public vision through mutual obedience to cultural norms—with the realm of unremitting light and watchful sight that Socratic thought propounds. The Apollonian calls inexorably upon the counterweight of Dionysian excess since it requires an 'outside' that it must harness for the internal coherence of social order to be possible. Socratism, in contrast, regards the Dionysian element of tragedy as an atrocity committed against the organizing precepts of reason.[86] Reason aims at nothing less than the universal illumination of the world by abstract technical principles which generate powerful auras of distance and 'objectivity'. From the rational-Socratic standpoint, the Dionysian is 'irrational' since it opens up cracks in the social order through which chaotic transgressions of organized space can flow in. Consequently, Socratism seeks to deny Dionysus. Where Attic tragedy invokes a productive conflict of visualizing perspectives which pits organized social order against a mythic imagination that draws on embodied drives, Socrates, armed with his 'great Cyclops eye',[87] sees in one flattened-out, saturated dimension and calls it 'totality' and 'truth'. Nietzsche's linking of Socratism with the primitive mythical Cyclops is deliberately provocative. It suggests that in the midst of his rationalism, Socrates's purportedly sophisticated and progressive viewpoint is as crude and backward as the most reviled of cave-dwelling life-forms depicted in Homeric epic. The grossness of the Cyclops, epitomized by his inability to see any other standpoint than his own, is paralleled by the abstract conceptual grossness of Socratic visuality. Socrates' perspective is one which crudely takes truth to consist of one all-embracing standpoint on the world. It is a perspective which denies all other perspectives rights unless they are compatible with its own.

The consequence of this outlook is a distinctive kind of violence. The Dionysian, on the one hand, is excised as mythical irrationality and error and cast out. The Apollonian, on the other, now stands opposed to the methodical light of a timeless 'reason'. Method is the unchanging order of rational rules. In contrast to this, the Apollonian world of daily life comes to be derided as pure error ('custom' is mere unreflective doing according to the dictates of convention, the world of 'belief' rather than true 'knowledge'). Socratism thereby embraces a visuality designed to obey the universal, ordering principle of a rationality which conceives of truth as the unchanging and pure concept of an intellectuality that exists sealed off from the contingent and inferior empirical realms of culture and law on one side and unruly nature on the other.

Consider, in relation to this last point, the famous discussion of the sun and the divided line offered by Plato in the *Republic* (507c–511e). Here the terms of Socratic visuality are starkly laid out. For Socratism, knowledge concerns what is true irrespective of mere belief. To know about beauty or morality, for example, involves comprehending the nature of beauty or morality in itself (i.e. something universal) rather than simply entertaining the belief that some particular thing is beautiful or some particular act is moral. Such knowledge concerns the eternal and unchanging nature of goodness, which is why 'philosophers are those who are capable of apprehending that which is permanent and unvarying'.[88] This ability is what makes philosophers the best rulers of any community since they will surely prove unswerving guardians of the truth contained within the community's laws and customs. However, even Socrates, it turns out, cannot actually tell us what goodness is. He has to resort to simile, substituting something in its place in order to convey what the good might be like. The simile concerns light and its source, the sun. Imagine a person who is using his or her sight to see something that is coloured: 'the sight will see nothing and the colours will remain unseen, surely, unless there is also present an extra third thing which is made specifically for this purpose'.[89] This extra 'something' is *light*. Light illuminates what is seen and thereby 'links the sense of sight and the ability to be seen'.[90] The source of light is one of the 'heavenly gods': the sun. The sun cannot be identified with the sense of sight or the eye. Rather, 'the eye's ability to see has been bestowed upon it and channelled into it, as it were, by the sun'—and one can think of the mind in the same way.[91] When the mind contemplates things that are illuminated by 'truth and reality', it has 'intelligent awareness and knowledge'. When, however, the mind contemplates things that are 'permeated with darkness (that is, when its object is something which is subject to generation and decay)', then its beliefs are 'less effective' because they are unstable, vague, and ill-defined.

It seems not to matter to Socrates that the simile of the sun can be taken as an admission that truth, too, remains vague and ill-defined. Indeed, his evasion of this concern is completed when the simile takes its place in a discursive metaphorics: the 'illumination' of the world is the immutable light of truth as the apprehension of timeless goodness—an account developed in the *Republic* in the parable of the cave.[92] Untruth, in contrast, is the 'darkness' of the changeable and finite. It is worth noting the point, so well made by Jacques Derrida, that here is an example of the 'heliocentrism' prevalent in Western discourse (not only in its metaphysics but its aesthetics and its politics). Light, in such a discourse, becomes the power of thought to survey a 'higher' reality than that vouchsafed by individuated daily life or the mere biological life of the body and so to penetrate into things themselves.[93]

The later Nietzsche will come to offer his own distinctive account of the cultural implications of Socratism, which we will turn to shortly. For the moment, one can proffer the following provisional argument. With Socratism, the Dionysian is subjugated and marginalized by a visualizing 'reason' in order to legitimize a form of conceptual domination.[94] One might say that the practices of discourse which rise from the social order are in this way turned on it. A form of domination thus emerges from the Apollonian realm of custom and tradition in the shape of an individuated Apollonian self that has separated itself from that domain and turns against it through an internalizing, neurotic intensification of the Apollonian calculative propensity to accede to the norm. Rule-following, which is a necessary condition of practical everyday life, becomes under the power of such a fixated gaze an abstracted conceptual fetish. Socrates, in other words, exemplifies a new kind of individuation: a kind of Apollonianism without reserve in which the locale of the *polis* and its citizenry is abandoned in favour of the standpoint of purportedly universality of rational categories of thought which adhere to faith in abstract calculative principles.[95]

The walls which enclose the *polis* and the normative barriers which regulate the Dionysian flows within it are thereby in their own turn enclosed within a rhetorical conceptual structure that claims for itself the right to exceed and contain them. The Socratic self no longer looks 'out' from the vantage point of protective Apollonian city walls onto the plane of unruly Dionysian exteriority. Instead, it proclaims possession of an absolute methodological perspective (a geo-intellectual territory) capable of looking down on Apollonian space and the Dionysian abyss beneath it in order to deny the reality of both, even as it occurs situated alongside them within the plane of their appearing. 'Reason' asserts the right to open up a larger, indeed 'universal', spatiality by way of an all-seeing vision that permeates everything because it claims to see abstractly—and abstraction is supposed to be equivalent to transcendence. Such reason craves the seeming clarity of number and law derived from the Apollonian realm, of procedural rules which endow action with certainty, but does so absolutely and so claims immunity from the ambiguities of the social and cultural world from which it springs. In this regard, the light of reason illuminates only in virtue of what it conceals from itself of itself. It is this illusion of detachment which renders of the world in all its cultural variety an object of manipulative thought.[96]

When subordinated to this Socratic demand, seeing and the practices associated with it become tools of 'representation' (objective 'picturing') put in service to the ruling power of conceptual abstraction. Reason becomes the source of the illumination of the totality of the environment, not a pragmatic consequence of the construction of a practical social world that endows selves with rational-social abilities in virtue of the particularities of time and place and their relation to the embodied environmental conditions on which they depend. 'Illumination' here stands for 'universal intelligibility' (a metaphor for 'Truth'). 'Vision' and 'light' become obedient similes of the intelligible. This visual articulation does not organize the realm of the seen in terms of the creative tension between Apollonian interiority and Dionysian exteriority, as tragic sensibility does. In Socratism all reference to externality and the tragic implications it contains is closed off in favour of an abundant and optimistic transcendence that encircles and saturates everything with the supposedly pure light of the concept.

Writing at the end of his productive life in *Twilight of the Idols*, Nietzsche returns to the topic of Socratic reason in a manner that allows for a further elaboration of the above points. The Apollonian mode is once more held to centre on provoking the eye, especially in such a way that it 'receives visionary power' which translates observation into expressive forms (painting, sculpture, epic poetry). Its celebratory quality is realized in the Homeric epic—as, one might add, is its defensive territoriality. Again, it is a matter of enclosing the environment within an ethos of order. The Dionysian, true to form, generates a contrastingly disruptive eruptive visuality. This is revealed as irresistible compulsion, a driven form of expression, in which transformation, movement, and mimesis (gesture) are concentrated in the entire sensory system of the body, which it pervades, announcing itself as 'the inability *not* to react'.[97] Here is the 'expression of the eternal joy of creation'.[98] In Socrates, in contrast, thought hurls itself onto reason in a manner suggestive of impalement. Socrates wants to make 'a tyrant out of *reason*'.[99] The light of Socratic rationality marks the emergence of a voyeuristic gaze designed to extirpate dangerous desires by subjecting them to an unending and eternal sleepless day from which there can be no hiding (and all the shame that this implies). What motivates Socrates is a fanatical, *un*reasonable desire to turn reason into an absolute, to endow it with a status of which it is not worthy and in this sense to do violent injustice to it. This desire is symptomatic of a commanding need whose perverse wish to 'establish permanent *daylight* [. . .] the daylight of reason' and embrace the unforgiving principle of '*daylight, rationality at all costs* [. . .] *instinct-free*'[100] reveals it to be in actuality far from rational.[101]

Reason embraced to the point of unreason . . . It has already been argued that Socratic reason is akin to a renegade Apollonian subjectivity that exchanges the primacy of city walls for the abstract visual power of the bloodless calculative concept, its aim being to make opaque walls, bodies, and anything else that lies beyond walls (exteriority) the passively transparent objects of manipulative thought. Socrates has thereby already managed what Horkheimer and Adorno, in *Dialectic of Enlightenment*, claim as the distinctive achievement of modern thought.[102] He has turned the world into 'a gigantic analytic judgment'[103] ripe for picking apart and putting to use in any way the modern (and not so modern) rationalistic desire for mastery sees fit. All that is needed for instrumentalism to emerge from its initial Platonic-Socratic articulation is for the avowed moral-metaphysical commitments of Socratism to be sloughed off. With this done, method alone is left standing as a bare, enclosing structure of rules and regulations which stake their claim to the right to regulate thought by dictating what is to count as meaningful and what not (a common name for this today is 'positivism'). The Socratic vision of the totality of the world already renders it a state of pure communicative connectedness, a structure of abstract rules amenable to the manipulations of a method-obsessed thinking. All its parts are conjoined to one another instantly by the omnipresent uniformity of the methodological, unifying principle of reason.

It is at this juncture that one can return to pondering the hypermodern. Reason, when it is conceived of as methodological illumination, is the ambition of a thought that would love to attain a pure connectivity through which every part can be instantly related to every other. It is this leap in abstract conceptual linking—the imagining of a thought as fast and penetrating as pure light, a 'pure reason' that determines everything's limits in terms of its own concept of 'intelligibility'—which is a necessary condition for the shaping of the regulated flows which characterize hypermodern saturated space. Virilio may have a powerful point when he traces modern technology's heritage back to religious and militaristic roots, but amongst these roots are also to be found the stems of burgeoning Western rationality, of a reason which is neither the timeless natural exemplar of human nature nor manipulation's or efficiency's instinctive enemy. Through the instant universal connectedness already present in the Socratic idealization of method and concept, the later instrumental-rational technological pursuit of ever greater communicative speed and efficiency is legitimized in a manner otherwise unthinkable: as the motorization of the world and the eroding and liquefaction of Apollonian certainties through the manipulation of Dionysian desires. In hypermodernity there thus persists a cleaving to the ancient Socratic dialectical faith that the technical working through of methodological principles necessarily yields not only correct results but fortuitous ones.[104] This is a positivistic faith. It is a faith which explodes with maximum force in the modern era when it is combined with and directed by the forms of power associated with contemporary technology. Within this mode of thought, practices are reformed and redirected. Visualization conforms neither to the demands of Apollonian tradition and convention nor to Dionysian bodily sensibility. Rather, the visible world is unveiled by an abstract visualization conducted under the banner of a 'method' according to whose dictates the intelligible is the universal principle which unifies the manipulable whole.

The roots of this instrumentalizing tendency reside in Socratic rational knowledge, in 'epistemology', with its justification of a space made visible as a thought-object and so rendered susceptible to being ruled by sovereign abstract conceptual power. Methodological thought, in this regard, claims the right to legislate; it becomes politics. What is Plato's 'philosopher king' if not a dislocated observer whose claim to authority springs from inhabiting the intellectual condition of distantiated abstraction? The ideal ruler, Socrates reminds us, will be a person who oversees city space by situating it within the totality already delineated by universal principles that transcend local principles. It does not matter if the ruler's authority is in the end ensured by the mythical invention of a 'noble lie' to gain common acceptance,[105] for this person has been taught to believe that once trained to orientate the 'vision of the soul' upward 'to look at the source of all light', he or she becomes the privileged possessor of a universal 'pattern for ordering their own life as well as that of society and the individual'.[106] The platonic ruler offers us the archaic pattern for the modern administrator pure and simple: it is a managerialist figure endowed with the keen 'rational' vision necessary to keep a sharp eye on all things.[107] Ruling is observation guided by abstract methodological principles which equate with a light that illuminates. The ruler-knower sees all in relation to the guiding principle which determines all, namely, the sovereign methodological principle of intelligibility.[108] It is this which endows rational light with the power that facilitates its emanation into the world of the visual.

In accordance with this demand, the ruler is, Socrates adds, akin to a good painter. The painter looks to the thing to be painted and continually refers back to it in order to create an accurate 'representation'. 'Truth', likewise, is representation guided by method. And faithful to the circular form of justification that was to show its face again unashamedly nearly two millennia later in the guise of logical positivism,[109] the representation of truth is the replication of the thing in accordance with the dictates of that method. In this, Socratic reason proffers the seeds of the less-welcome Enlightenment tendency to conceive of tradition as a condition of error and backwardness standing in need of correction and administrative harmonization ('rationalization', making efficient by a process of constant revision) according to authoritative conceptual principles of correspondence with a 'higher' plane of vision. Whatever does not 'correspond' to methodological reason lacks truth and should beware . . .

Understood in these terms, archaic reason is what from the outset introduces the possibility of the obsolescence that erupts with full force within the dominant logic of modernity and its liquefying hypermodern variant. The outdated domains of tradition and biological inheritance (Apollonian and Dionysian elements) must either accord with or yield to the continual and newly (re)discovered rule of a universal rational demand that opens the bright path to the future. The spheres of custom and convention and the body alike are condemned to obsolescence as soon as one begins to refer them back (as Socratism does) to the level of an autonomous knowledge that is, as Hobbes says,[110] simultaneously older (in the sense of being rooted in higher enduring principles that have always and everywhere been the case) and younger (in the sense that the arrival of this enduring higher knowledge is marked by its usurpation of tradition and inheritance). In this double movement of thought lie the seeds of the presentism which, as Lipovetsky notes, distinguishes the eruption of the condition of hypermodernization. Hypermodernity thereby announces itself, as Virilio's account of the urban suggests, in a peculiar and violent coming together and circulation of ancient tendencies and burgeoning modern forms.

This coming together, however, need not be conceived of as an act of vandalism directed against an unhistorical human nature, as Virilio is committed to holding to be the case. As has been said, personhood is perfectly amenable to being conceived of as an unwitting cultural achievement whose origins need not be conceived of in thrall to any trans-historical authority. The historically cultivated Apollonian subject's Dionysian embodiment is already waiting for violence to be done to it. The archaic, one might say, in this way juts into the modern precisely because it has given rise to it by way of the combination of practices and struggles over meanings which make up its history. It persists, moreover, not as an outmoded, dead remnant that 'progressive' modernity has somehow left behind. The presence of the archaic proclaims the myth of 'progress' in this sense. The faster it goes, i.e. the more hyper-intense it becomes, the more the modern churns and drags up the ancient, cloying residues of its own self-creation on which it floats. For, it is fair to say, modernity has no 'foundation' any more than the bunker logic of defence which permeates it does. The archaic inheres as a living and circulating component of the present itself. This inherence is not simply, as Virilio argues, the archaism of an ever returning violence linked to the watchful defensive requirements of the city. It is in no small part that. But it is also what the city needs and must repress: the unruly Dionysian luminescence of the body, of an exteriority that transcends lawful observance and has a propensity to rise up in dangerous ecstasy. It is, in addition, the visuality of an ancient universalism endowed with a tendency to colonization through a rational illumination that chains things together: light as thought, thought as image, image as control, control as communication, communication as an interlinked and coordinated 'representation' of the environment arranged in a simultaneity that renders it unified and so pliable to the 'will' of the knower. The logic of hypermodernization is not related to one thing. It is not one condition, one way of seeing but a concatenation of Apollonian, Dionysian, and Socratic elements with the latter, in the guise of the instrument of method, forever seeking to enclose the other two.

Speaking like Bauman, one can say that this method as it encloses liquefies the very Apollonian space of tradition it opens out of and depends upon in order to be what it is (*one* word for this is 'globalization'). In that regard, one might think of hypermodernization as exemplifying both the logic of enclosure associated with the bunker and that logic's unwitting self-dissolution. Nothing can aspire to permeate everything without thereby eroding itself in unpredictable ways in the process. To the extent that method aspires to be 'like' nature, the modern technological expression of method might, in this regard, be unwittingly returning us to nature. The bunker of Western culture might well flip over and collapse at the hands of the elements it seeks blithely to master.

SECTION 5: PRELUDE TO PART 2

It has been argued above that within the modern era there arises an increasingly saturated social environment permeated by an industrialized suffusion of space. An initial tentative analysis of a technological artefact like the photograph of the bunker structure reveals this saturation. Its essence resides in its being an unfurling of the potential of this observational-military articulation of spatiality which from the outset typifies the domain of the city as a dromological environment. As has also been suggested, what ought to be striking about such a reading is its conjoining of two apparently disparate elements already voiced in various ways by figures located at different points in the history of the opening out of modernity and the Enlightenment. On the one hand, one can say that modern urbanization and the visual technologies which exemplify it reveal a disruption of existing social forms which produces an overwhelming sense of the supplanting of tradition within an endlessly replicating ethos of industrialized, cosmopolitan newness. The modern, when considered in this way, is the domain of what is most contemporary (of the 'cutting edge', to use a common and unimaginative, managerialist phrase).

The notion of 'fashion' as exemplified in many contemporary products is one example of the mode in which the modern is expressed with all its eye-catching visuality pressed to the fore. At the same time, modern space and the elements which facilitate its articulation hark back to something of a seemingly altogether different order, namely, the primitive demand according to which defensive considerations take precedence over all others. Such a demand walled up the ancient space of the *polis* and in doing so delineated its environs as a domain of observation. To the extent that military sensibility (the Apollonian need for security) persists in shaping the accelerating modern environment, such a space, however 'modern' in the sense of being marked by the optimism of relentless renovation and newness, is at the same time tinged with a quality of primitive fear and watchfulness, a quality evident in Clausewitz's thinking when he outlines the Napoleonic origins of the modern manifestation of total war. In contemporary culture, this duality of qualities shows itself in surveillance technology and in the fear of terror facilitated by the flood of visual communications systems which can relay violence smoothly and easily. The ancient condition of total war comes in this way to make its presence felt even in the peacetime city environment. The ongoing condition of this divided dynamic and the diverse and chaotic currents it generates typifies a saturated modality of seeing which can be deemed quintessentially hypermodern.

One might, following Giorgio Agamben,[111] argue that this dual movement is the West's most spontaneous political gesture. It is linked to a warlike, 'dromocratic' tendency to seek speed, for the most effective form of countermeasure when it comes to threats is to be able to respond to them with a greater rapidity than they possess (which ideally means cultivating the predictive foresight and technical means to prevent them before they happen). It is also conjoined with the Socratic desire for a transcendence achieved by the transparency of enlightened 'rational' vision. In this sense, the city wall is extended and elaborated into a complex metaphor by observational regimes and, later, abstract thought in conjunction with optical and then electronic visual technologies. What would it mean to 'see' through this model? How might one visualize the dissolution of the bunker? Could one? Another snapshot?

Here is something 'captured' in the process of becoming archaeology. Two years later, the flow of the sea had overrun its frame altogether.

How might one seek to 'see into' the hypermodernizing mode of 'transparency' which has here been allied with Socratic thought and the visuality it cultivates as the epitome of truth without either replicating it or seeing right through it? Certainly, given the account offered above, this is not to be attempted by cleaving unreflectively to the dictates of 'method' in the certain faith that it will see one (a)right. A more open-ended approach seems to be demanded. This is what is offered in the chapters below which make up part 2 of this text.

NOTES

1. One can, of course, 'see' the sun by projecting its light through a telescope onto a surface or gazing at it through a thick neutral density filter. In such cases, however, one 'sees' in relation to the equipment in virtue of which the intensity of sunlight is rendered amenable to being translated into something else which, in these examples, is either through an act of projection or reduction. In the latter case (reduction), one 'sees' the filter rather than the 'light'; in the former, one 'sees' the combination of glass and tube in relation to its peculiar propensity to channel waves.

2. See Paul Virilio, *The Art of the Motor*, trans. Julie Rose (Minneapolis: University of Minnesota Press, 1995). Virilio regards Nietzsche as a thinker of the 'hype' associated with the ambition of achieving a technologically perfected humanity able to dwell in a 'resplendent beyond' (108). He also identifies Nietzsche with a gross, Pelagian, modernist, technological fantasy of a post-human world populated by a 'superspecies' and as a eugenically inclined thinker who cultivates the 'myth of great health' (129). Such ideas harmonize perfectly well with contemporary '*public health ideology*, an ideology that is no longer particularly interested in maintaining public health so much as in continually improving it' (122), which means cultivating a population amenable to ingesting (and, of course, therefore purchasing) chemically enhanced forms of nutrition so that they become ever more like the animals that up to now have been refined by breeding for domestic purposes. As if this were not enough, Nietzsche is for Virilio also the quintessential thinker of modern secularism in its most brutal visage: 'The death of God—I'm speaking of Nietzsche—is an abomination of desolation in recent political history [. . .] After Nietzsche we went away from doing away with God to doing away with the question of God [. . . to . . .] the evacuation of God' (Paul Virilio and Sylvère Lotringer, *Pure War*, trans. Mark Polizzotti (Los Angeles, CA: Semiotext(e), 2008), 141–42). Virilio's reading of Nietzsche's views on health is debatable. Nietzsche does not conceive of health in these essentially biopolitical terms. Indeed, he explicitly questions the very notion of the existence of a 'health as such'. For a discussion of his complex, and indeed anti-biopolitical, attitude to health, see Peter R. Sedgwick, 'Nietzsche, Illness and the Body's Quest for Narrative', *Health Care Analysis* 264 (2013).

3. Arthur Kroker, *The Possessed Individual* (London: Macmillan, 1992), 28.

4. See Martin Jay, *Downcast Eyes: The Denigration of Vision in Twentieth-Century French Thought* (Berkeley: University of California Press, 1994). According to Jay, twentieth-century French thought is both highly sensitive to the visual 'and no less suspicious of its implications' (588). In effect, Jay says, modern French thought, from Bergson to Bataille to Sartre, Derrida, and Foucault, is characterized by a loss of faith in ocular perception. Virilio in fact only receives two mentions in Jay's book, both of

them in footnotes. Perhaps this is because he is, as Arthur Kroker comments, 'so clearly opposed to most of contemporary French thought' (*The Possessed Individual*, 38).

5. Paul Virilio, 'The Illusions of Zero Time', in *Virilio and Visual Culture*, ed. John Armitage and Ryan Bishop, trans. Chris Turner (Edinburgh: Edinburgh University Press, 2013), 28–36.

6. John Armitage and Ryan Bishop, 'Aesthetics, Vision and Speed: An Introduction to Virilio and Visual Culture', in *Virilio and Visual Culture*, 1–27.

7. For a useful account of the development of Virilio's view of the bunker form, see Mike Gane, 'Paul Virilio's Bunker Theorizing', in *Theory, Culture and Society* 16, nos. 5–6 (1999): 85–102. Gane notes that Virilio's early thinking, which was in no small part a product of his working with the architect Claude Parent, was characterized by an attempt to develop the bunker structure in the pursuit of radical and transformational ends: 'to make military architecture, as it condensed social energy and revealed new functions, serve other functions' which could revolutionize '"static space" into a habitat "mobilized by the function of the oblique" and liquid technology' (87–88). Parent's notion of the 'oblique function', a tilting of the ground of a structure, is intended to undo the accepted notion of the wall as a vertical structure. The effect of this is to transform the everyday and casually accepted experience of constructed space insofar as a person moving through an obliquely inclined environment is forced to traverse it in an upward or downward movement, bringing the sense of that space to the fore. The most notable result of Virilio and Parent's collaboration is the church of Sante Bernadette du Banlay in Nevers, France. The exterior of the church is strikingly modelled on the concrete defensive bunkers constructed by the Germans during World War II which litter the European littoral. The first fruits of Virilio's radical re-theorization of the bunker form are presented in the text *Bunker Archaeology* (discussed below). For a useful discussion of Virilio and architecture, see Neil Leach, 'Virilio and Architecture', in *Theory, Culture and Society* 16, nos. 5–6 (1999): 71–84.

8. Eric Hobsbawm, *Age of Extremes: The Short Twentieth Century* (London: Abacus, 1995). Hobsbawm designates this period as extending from 'the years from the outbreak of the First World War to the collapse of the USSR', i.e. 1914–1991. The name 'Short Century' captures well the sense of brevity and fluidity which permeates it. The fluidity is demonstrated not only by the Short Century's being typified by a 'moral crisis' which attacked 'the assumptions of modern civilization' but the simultaneous corrosion of 'the historic structures of human relations which modern society inherited from a pre-industrial and pre-capitalist past' which 'enabled it to function' (11). One might thus deem the Short Twentieth Century as initiating a crisis of self-dissolution. It is also, of course, the age of a hitherto unparalleled mechanization of war.

9. This is not to deny that the modern defensive bunker, which takes memorable form in the diminutive 'pillbox' design erected all over Britain during World War II, has a long heritage. Mike Osborne notes that its ancestry stretches from the Renaissance 'caponier', a defensive structure overlooking the ditch surrounding a fortress, to the Napoleonic *tour-modèle* and 'blockhouses' built during the Boer War. See Mike Osborne, *Defending Britain: Twentieth-Century Military Structures in the Landscape* (Stroud: History Press, 2010), 244ff. See also Bernard Lowry, *British*

Home Defences, 1940–45 (Oxford: Osprey, 2004), 15ff. A sketch for a caponier by artist Albrecht Dürer (1471–1528) is held by the U.S. Library of Congress Rare Book and Special Collections Division and can be viewed online (www.loc.gov/pictures/item/2006690453/). The bunker reaches its apogee in the defensive constructions built across Europe during and between World War I and II. The most famous of these is probably the Siegfried Line, whose construction was ordered by Field Marshall von Hindenburg in September 1916. Initially, this was intended as a 'semi-permanent' defensive line to be constructed five to thirty miles behind the front' (Martin Gilbert, *First World War* (London: Harper Collins, 1995), 288).

10. Paul Virilio, *Bunker Archaeology*, trans. George Collins (Princeton, NJ: Princeton University Press, 1994), 45.
11. Virilio, *Bunker Archaeology*, 44.
12. Günther Anders, 'On Promethean Shame', in Christopher John Müller, *Prometheanism: Technology, Digital Culture and Human Obsolescence* (London: Rowman & Littlefield, 2016), 30.
13. Virilio, *Bunker Archaeology*, 44.
14. 'Impossible' because no one, except by recourse to fantasy and metaphor, can dive into a photograph, its two dimensions folding out before them into the lived domain of three (one says 'two' and 'three' here because the fourth, temporality, is a condition of both).
15. Paul Virilio, *The Vision Machine*, trans. Julie Rose (Bloomington: Indiana University Press, 1994), 13.
16. Vilém Flusser, *Towards a Philosophy of Photography*, trans. Anthony Mathews (London: Reaktion, 2006), 10.
17. David Wills, 'Raw War: Technotropological Effects of a Divided Front', *Oxford Literary Review* 31, no. 2 (2009): 133–52.
18. Virilio, *The Vision Machine*, 14.
19. Virilio, *The Vision Machine*, 14.
20. Paul Virilio, *Open Sky*, trans. Julie Rose (London: Verso, 2008), 89.
21. The political implications of this are concerning. In Verena Andermatt Conley's words, 'Virilio cautions that the totalitarianism of the information-medium might be even more powerful than the traditional political totalitarianism of the old national-socialist or communist hues' ('The Passenger: Paul Virilio and Feminism', in *Theory, Culture and Society* 16, nos. 5–6 (1999): 201–14).
22. John Berger, *Ways of Seeing* (London: Penguin Books, 2008), 139.
23. See Paul Virilio, *Speed and Politics: An Essay on Dromology*, trans. Mark Polizzotti (Los Angeles: Semiotext(e), 2007), 29–48.
24. See Adam Smith, *The Wealth of Nations*, ed. Andrew Skinner (London: Penguin, 1999), bk. 3, chap. 3:496ff for Smith's view that liberty originates in the city and is subsequently dispersed into the country.
25. Smith, *The Wealth of Nations*, 115–16.
26. John Berger, *About Looking* (London: Bloomsbury, 2009), 104.
27. Virilio, *Speed and Politics*, 31.
28. See Martin Heidegger, *Holzwege* [Wood-paths], trans. Julian Young and Kenneth Haynes (Cambridge: Cambridge University Press, 2002).

29. Friedrich Kittler, *The Truth of the Technological World: Essays on the Genealogy of Presence*, trans. Erik Butler (Stanford, CA: Stanford University Press, 2013), 148.

30. Virilio, *Speed and Politics*, 31.

31. As Friedrich Kittler comments, Virilio's thought makes a decisive link between the rise of optical forms, militarization, and the strategic use of information, namely, that 'the strategic interest in faster information [. . .] crucially accelerated the explosive rise of optical media over the last hundred years' (*Optical Media*, trans. Anthony Enns (Cambridge: Polity 2010), 42).

32. Virilio, *Speed and Politics*, 70.

33. See Virilio, *The Art of the Motor*, 151ff.

34. A point made by Jeff Noonan, who notes the links between acceleration and recent political developments such as the UK Brexit vote of 2016. See 'Paul Virilio and the Temporal Conditions of Philosophical Thinking', *Time and Society* (2017): 1–20. Noonan self-consciously presents Virilio as a kind of Old Testament 'prophet of doom' in this regard: What is at stake in modern acceleration is nothing less than 'a loss of our humanity' (19).

35. This shows in exemplary fashion the way in which following Virilio's line of thought 'encourages us to rethink visuality as an active, embodied practice that does not differentiate between the ability to see visual technologies and the constitution of lived experience' (Rob Bartram, 'Visuality, Dromology and Time-Compression: Paul Virilio's New Occularcentrism', *Time and Society* 13, nos. 2–3 (2004): 297.

36. And of course, if need be, responded to with further more direct strategies of containment and extirpation.

37. Powerful testimony to this in relation to the post-war era was made in the famous 1961 Military Industrial Complex speech of U.S. President Dwight D. Eisenhower. The speech identifies the emergence of an 'immense military establishment and a large arms industry' which threatens to exert a decisive influence, shaping the future 'economic, political and even spiritual' aspects of civilian life. See Dwight D. Eisenhower, 'The Military-Industrial Complex', in *21 Speeches That Changed Our World*, ed. Chris Abbot (London: Random House, 2012), sec. 4.

38. In the UK alone, it is thought that as many as 28,000 defensive bunkers were built, many of which survive in varying conditions of decay. See Osborne, *Defending Britain*, 7.

39. For a discussion of the gated community, see Edward J. Blakely and Mary Gail Snyder, *Fortress America: Gated Communities in the United States* (Washington, DC: Brookings Institution Press, 1999). Blakely and Snyder date the earliest of such communities to the period of the Roman invasion of Celtic Britain (that is, a Britain in which there was no 'English' population) around 300 BC (3–4). Interestingly, they comment that 'contrary to popular belief, the walls around these settlements were seldom to protect against external invaders but rather to guard against the local villagers [. . .] Later, fortresses also served to protect against invaders or internal warring factions' (4). The fortress thus emerges as a protective measure designed to regulate and thereby limit localised threats that accompany the movements of population and goods to the dominant power.

40. 'Towers are vertical bunkers' (Virilio and Lotringer, *Pure War*, 211).

41. One might want to consider the fact that surveillance cameras as a rule tend to record movement. Their effectiveness, however, turns on the fact that they record a sequence of more or less continuous movements which can be returned to in order to be frozen at any point. The 'freeze frame' testifies to the surveillance camera's kinship to the still camera and reminds us that all 'cinematic' representations of movement are in fact strings of still images flashing before the eye at the speed of around twenty-four frames per second.

42. For a discussion of these imaging systems, see Laura Kurgan, *Close Up at a Distance: Mapping, Technology, and Politics* (New York: Zone Books, 2013). Coincidentally, an early example of the spirit of this kind of geo-mapping is offered in the shape of the much-loved London Underground Map designed by Harry Beck in 1933. The map articulates the relation between stations in terms of the purely structural condition of connectedness—a mode of connection in this case based on that, at the time, most up-to-date of technological patterns, the electrical circuit diagram—which usurps the geographical specificity of landscape. See https://tfl.gov.uk/corporate/about-tfl/culture-and-heritage/art-and-design/harry-becks-tube-map.

43. For a discussion of this, see Sigrid Weigel, 'The Flash of Knowledge and the Temporality of Images: Walter Benjamin's Image-Based Epistemology and Its Precondition in Visual Arts and Media History', *Critical Inquiry* 41, no. 2 (2015): 344–66.

44. As Anne Friedberg comments, for Virilio the screen needs to be understood in 'a metaphoric register, [as] a virtual surface which overrides any specificities of its media-formation' ('Virilio's Screen: The Work of Metaphor in the Age of Technological Convergence', *Journal of Visual* Culture 3, no. 2:184.

45. This is the domain of the social order which denudes itself of night and dark. It is a world of permanent illumination and sleeplessness, 'a time without time, a time extracted from any material or identifiable demarcations, a time without sequence or recurrence' (Jonathan Crary, *24/7* (London: Verso, 2014), 29). Crary links the emergence of this world to the rise of modern capitalism which, because of its emphasis on 'accumulation, financialization, and waste', is driven to destroy 'everything once held in common' (128). The contemporary city's relation to sleep is well captured by the recent appearance of two 'sleep companies' in New York: Dreamery and Nap York. These offer New Yorkers the opportunity to take a brief rest in 'nap pods' as a respite from hectic city life. Lunchtimes are busiest. See Anne Kadet, 'Metro-Money: Napping in a City That Never Sleeps', *Wall Street Journal*, 8 August 2018.

46. For an illuminating discussion of urbanization and terror, see Julian Reid, 'Rethinking Relations Between War, Modernity and City Spaces after 9/11 ', *Space and Culture* 7, no. 4 (November 2004): 396–408.

47. Terrorist attacks, such as the notorious murderous onslaughts on the Twin Towers in New York in 2001 or the coordinated assaults in Paris culminating in the slaughter at the Bataclan Theatre, exemplify this metro-political shift. See Virilio and Lotringer, *Pure War*, 209–10.

48. See Thucydides, *The Peloponnesian War*, trans. Martin Hammond (Oxford: Oxford University Press, 2009). The text relates how the expeditions of this near three-

decade-long war between Athens and Sparta war dragged in entire populations. Vast expenditure was incurred by the Athenian state and private individuals alike on the vast armies and sea fleets necessitated by the conflict. It also initiated the movement of population, livestock, and goods from the country into the *polis*, implicating the entire community and all it owned in the conflict. The long-term consequence of this extended and all-encompassing strife was, unsurprisingly, the decline of Athenian power.

49. See Carl von Clausewitz, *On War*, trans. O. J. Matthijs Jolles, in *The Book of War* (New York: Modern Library, 2000), 914–16. Clausewitz was a seasoned veteran of the European conflicts of the 1790s and early 1800s. He was also the originator of the maxim 'War is the continuation of politics by other means'. For Clausewitz, by the eighteenth century European war had undergone a process of restriction. Financial and strategic limitations on all sides meant that the attractions of entering into escalating conflicts were mitigated and war had for this reason become 'more an affair of the government' than of the populace (915). War was thereby denuded of its more 'extreme' possibilities with the result that it came to mean 'only diplomacy somewhat intensified' (914), consisting for the most part of small-scale conflicts over strategically valuable pieces of land, fortresses, and the like. However, in the 1790s, Clausewitz argues, the initiation of hostilities between Britain and post-revolutionary France transformed war. It brought about the revelation of a new form of military power, the French Revolutionary war machine. With this war machine, an entity of 'such a force as no one had had any conception of made its appearance. War had again suddenly become an affair of the people, and that of a people numbering thirty millions, every one of whom regarded himself as a citizen of the state' (916). With this was initiated the modern era of war without restriction in which the population of a nation could be co-opted into struggles seemingly devoid of limit. All that such an engine of war required to accelerate to devastating and unstoppable proportions, Clausewitz adds, was the requisite organisational ability. This was soon supplied in the person of Napoleon, who pounced on the potential of the revolutionary-industrial cultural legacy, turning its force on Europe to devastating effect in the Great War (a title later usurped by World War I) which tore across Europe between 1803 and 1815. With Napoleon, the link between war and a burgeoning industrial urban terrain was made explicit as the industrialised space of 'revolutionary' movements of all kinds was unleashed on the European landscape. Where Napoleon triumphed, the rest followed so that by 1814 all Europe had adopted the strategy of making war 'an affair of the people' (917), doubling the sizes of their armies and focusing all the economic and cultural forces of the emerging, modern nation state to that end.

50. See David Gates, *The Napoleonic Wars, 1803–15* (London: Pimlico, 2003), especially chapter 7, 147ff. In the Napoleonic conflict, Gates notes, the industrialized British economy was confronted by a largely agricultural French one 'still at the proto-industrialization level of development'. The latter, however, hoped to bring 'its superior resources and population' to bear on its 'relatively tiny' foe (156). It is this strategy which marks the initiation of the turn to total war. The devastation entailed by the conflicts is made clear enough when one considers the estimated 5 million people believed to have perished in them (270ff).

51. For an extensive engagement with the various interpretations of the meaning of 'total war', see Manfred F. Boemke, Roger Chickering, and Stig Förster, eds., *Anticipating Total War: The German and American Experiences, 1871–1914* (Cambridge: Cambridge University Press 1999); Roger Chickering and Stig Förster, eds., *Great War, Total War: Combat and Mobilization on the Western Front, 1914–1918* (Cambridge: Cambridge University Press, 2000); Roger Chickering and Stig Förster, eds., *The Shadows of Total War: Europe, East Asia, and the United States, 1919–1939* (Cambridge: Cambridge University Press 2003); and Roger Chickering, Stig Förster, and Bernd Grenier, eds., *A World at Total War: Global Conflict and the Politics of Destruction, 1937–1945* (Cambridge: Cambridge University Press 2005).

52. See the British government leaflet 'Air Raid Precautions: Directions for the Erection and Sinking of the Galvanised Corrugated Steel Shelter' (London: UK Home Office, 1939).

53. See, for example, Jason Burke, 'How Can Cities Protect Their Citizens Against Vehicle Attacks?', *Guardian*, 10 August 2018.

54. See www.knightscope.com/demo. This company manufactures mobile security robots designed to patrol 'corporate campuses, manufacturing plants, malls, logistic facilities, hospitals, stadiums, warehouses, apartment complexes, and single family home subdivisions across the United States'.

55. See https://efus.eu and http://media4sec.eu. Efus was established in 1987 under the auspices of the Council of Europe. The media4sec project seeks to 'build a community of stakeholders engaged in using social media for public security' in order to 'enhance the use of this tool' (http: //media4sec/about).

56. That is, from iconic structures which endow a particular sense of place, such as the Acropolis, St. Paul's Cathedral, the Eiffel Tower, the Temple of Amritsar (whose shape the bunker appears to mimic), the Jameh Mosque, and so forth.

57. Virilio, *Bunker Archaeology*, 45.

58. Virilio, *Bunker Archaeology*, 44.

59. Virilio, *Speed and Politics*, 86.

60. James Der Derian, 'The Conceptual Cosmology of Paul Virilio', *Theory, Culture and Society* 16, nos. 5–6 (1999): 224.

61. 'Humanity cannot be improved' (Paul Virilio, *Politics of the Very Worst*, ed. Sylvère Lotringer, trans. Michael Cavaliere (New York: Semiotext(e), 1999), 88).

62. See Sean Cubitt, 'Virilio and the New Media', *Theory, Culture and Society* 16, nos. 5–6 (1999): 127.

63. Ian James, 'Phenomenology in Diaspora: Paul Virilio and the Question of Technology', *French Cultural Studies* 17, no. 3 (2006): 330. For James, this pathos revolves around the 'nostalgia for a lost immediacy or possibility of presence [which] articulates an unresolved tension in Virilio's work'. As is apparent in what follows, I do not seek to resolve the tension, simply to sidestep it.

64. Scott McQuire, 'Blinded by the (Speed) of Light', *Theory, Culture and Society* 16, nos. 5–6 (1999): 153.

65. Thus, according to Douglas Kellner, there is a tendency in Virilio to blame technology for contemporary woes rather than the sociopolitical forces of capital

which drive technology ('Virilio, War and Technology: Some Critical Reflections', *Theory, Culture and Society* 16, nos. 5–6 (1999): 120.

66. Paul Virilio and John Armitage, 'From Modernism to Hypermodernism and Beyond: Interview with John Armitage', *Culture, Theory and Society* 16, nos. 5–6 (October–December, 1999): 36.

67. See Bernard Stiegler, *Technics and Time, 1: The Fault of Epithemus*, trans. Richard Beardsworth and George Collins (Stanford, CA: Stanford University Press, 1998), 134ff. It should be added that arguments have been made to the effect that Stiegler's own approach replicates 'the traditional architecture of anthropology' that it seeks to break with, i.e. 'the opposition between the living and the non-living, the natural and the cultural and [. . .] the human and animal' (Arthur Bradley, 'Originary Technicity? Technology and Anthropology', in *Technicity*, ed. Arthur Bradley and Louis Armand (Prague: Litteraria Pragensia, 2006), 96). David Wills has developed an interesting account of originary technicity which draws in part on Stiegler (*Dorsality* (Minneapolis: University of Minnesota Press, 2008)). Wills calls this the 'dorsal turn' since it occurs invisibly, behind the back. For Wills, language is the foremost 'technological system' (15) which forms the paradigm for this account. Language, in other words, is always already technologized.

68. Sean Cubitt identifies what he regards as a significant problem here. Since Virilio 'believes in a wholeness of perception independent of cognition' and 'since he does not believe that perception is in every case mediated [. . .] he understands mediation as a fall from grace, from a natural perception which, in its integration of human and world, provided the basis of human identity' ('Virilio and the New Media', 135). As is probably clear from the outset of this text, cultural mediation is endorsed as a 'natural' condition of personhood.

69. Virilio and Lotringer, *Pure War*, 214.

70. Friedrich Nietzsche, *The Birth of Tragedy*, trans. Walter Kaufmann, in *Basic Writings of Nietzsche* (New York: Basic Books, 1968).

71. Kittler, *The Truth of the Technological World*, 20.

72. Nietzsche, *The Birth of Tragedy*, sec. 2.

73. Nietzsche, *The Birth of Tragedy*, sec. 1ff.

74. As Tracy B. Strong notes, the problem of tragedy for Nietzsche is 'a political problem' which turns on 'the task of a creative renewal of a collective identity'. Nietzsche's position in effect turns to aesthetics rather than politics as providing the 'groundwork' for this renewal ('Nietzsche's Political Aesthetics', in *Nietzsche's New Seas*, ed. Michael Gillespie and Tracy B. Strong (Chicago: University of Chicago Press, 1988), 161–62).

75. Friedrich Nietzsche, *On the Genealogy of Morality*, ed. Keith Ansell Pearson, trans. Carol Diethe (Cambridge: Cambridge University Press, 2017), bk. 2, sec. 2. Needless to say, this sense of autonomy, of being the possessor of an unyielding '"free" will', is an illusion.

76. The 'individual' here denotes a condition of personhood such that someone endowed with self-understanding interprets his or her environment from the standpoint of their place in a world (i.e. a domain of meaningfulness, a culture). Nietzsche

is committed to the view that there is no sense in the notion of an 'individual' person independently of this condition. There can, of course, be individual instances of creatures: a herd of zebra is composed of individuals in this sense, but these individuals are not individuated in the sense of being persons.

77. Nietzsche, *The Birth of Tragedy*, sec. 1.

78. Nietzsche, *The Birth of Tragedy*, sec. 5.

79. Kittler, *The Truth of the Technological World*, 21.

80. Nietzsche, *The Birth of Tragedy*, sec. 8. Nietzsche notes that the Greek theatre's structure, which consisted of a terrace of 'concentric arcs', meant that everyone present was able to '*overlook*', in the sense of both observe and ignore, the cultural environment that surrounded them and thus become another member of the Dionysian throng of the chorus.

81. One might notice that this visual aspect compromises the terms of the Apollonian-Dionysian duality that Nietzsche sets in place in *The Birth of Tragedy*. It does. This does not, however, compromise the reading being proffered here.

82. The logic of this overturning is analysed insightfully by Mikhail Bakhtin in his account of the carnivalesque. Bakhtin associates it with medieval anarchic festive celebrations (the Feast of Fools, etc.) in which the hierarchy of social order was temporarily and symbolically ridiculed (*Rabelais and His World*, trans. Hélène Iswolsky (Bloomington: Indiana University Press, 2009)).

83. Nietzsche, *The Birth of Tragedy*, sec. 8.

84. For the individuated self, the body becomes an object of observation. This is the case for one's own body and the bodies of others. Touch is in this way separated as one of the 'faculties' of the body (to see, to hear, to smell, to touch, to taste). The body, in other words, is divided up into a series of functions just as the world of social order individuates by identifying and so dividing 'selves' up into distinctive social roles and their associated functions.

85. Nietzsche, *The Birth of Tragedy*, sec. 19.

86. Benjamin Bennett argues that *The Birth of Tragedy* is in fact a decisively Socratic text. See his essay 'Bridge: Against Nothing', in *Nietzsche and the Feminine*, ed. Peter J. Burgard (Charlottesville: University of Virginia Press, 1994), 289–315. It is 'constructed on a thoroughly organized and entirely decipherable systematic basis [. . .] that fast approaches being a *summary* of Western philosophical systems' (298). This systematic foundation is 'unquestionably Socratic in character' in its arrogant ambition to master reality by way of 'an objectively valid reason' (289). The reader is in turn asked to insert him- or herself in the text in the form of a Socratic believer and then overcome this prejudice by learning to manipulate their 'subjection' to this dominant model of thought 'affirmatively' (300). This attempt to free the reader immanently is, Bennett argues, compromised because the text of *The Birth of Tragedy* still imposes 'a specific role on the reader and so compromises its own revolutionary force by erecting a kind of bridge' which replicates the structure of the authority it claims to overthrow. Such a reading depends on acceding to the idea that the structure and argumentation of *The Birth of Tragedy* is in effect Socratic through and through. As ought to be evident from the reading proffered here, I do not endorse this view.

The text may be situated within the logic of Socratism, but it invokes modalities of seeing which overturn the authority of Socratic reason's unifying fetishism.

87. Nietzsche, *The Birth of Tragedy*, sec. 14.

88. Plato, *Republic*, trans. Desmond Lee (London: Penguin, 2007), 484b.

89. Plato, *Republic*, 507d–e.

90. Plato, *Republic*, 507c–508a.

91. Plato, *Republic*, 508b, 508d.

92. See Plato, *Republic*, 514aff.

93. See Jacques Derrida, 'White Mythology', in *Margins of Philosophy*, trans. Alan Bass (London: Harvester Wheatsheaf, 1982), 209–71, especially 248ff.

94. Other interpretations are of course possible. The most thought-provoking is offered by Stanley Rosen in *The Question of Being: A Reversal of Heidegger* (New Haven, CT: Yale University Press, 1993). Rosen holds that Platonic 'Ideas are not intended as topics of metaphysical or ontological analysis leading to a systematic theory or account of the structure of the whole' for the

> 'Idea is not a detached original in the sense of a model that is copied by the image. A dog is not an image of the Idea in the sense that a photograph of a dog is an image of the living dog. If this were so, the original itself would have the look of a dog, and we would again have to distinguish between the "original" and its look. If the Idea provides the dog with its look, it must do so in some way other than by cognitive copying. The Idea [rather] *is* the look.' (59)

In effect, Rosen holds Plato's theory of Forms to not be imagistic but to function as a kind of meta-rule. The Idea provides the rule which governs the gaze as it looks. On Rosen's reading, for Plato/Socrates the doing precedes the engagement with entities. Thus we are endowed with a practical knowledge of action with regard to how to satisfy our needs independently of any particular entities we subsequently come across that meet the criteria of satisfaction. This is a kind of spontaneous knowledge which springs from our being 'immersed within the everyday' (70). Rosen's persuasive account does not, however, engage with the heliocentric metaphorics at work in Socratism which is at issue here.

95. Socrates, one should note, is not the obedient observer and defender of the rules of the social bond. On the contrary, he proclaims himself the independent bearer of the universal light of a timeless reason that accords with nature, superior to Apollonian illusion and Dionysian luminescence alike.

96. Witness, by way of illustration, the thought experiment enacted in Plato's *Republic* of imposing the rule of philosophical reason on the world in the form of an idealized *polis*. This is a polis in which everything is governed by the commands of rational dictates which gain their authority from an unchanging moral truth.

97. Friedrich Nietzsche, *Twilight of the Idols*, trans. Duncan Large (Oxford: Oxford University Press, 1998), 10.

98. Nietzsche, *Twilight of the Idols*, 4.

99. Nietzsche, *Twilight of the Idols*, 10.

100. Nietzsche, *Twilight of the Idols*, 11.

101. One only worships a condition of permanent daylight, Nietzsche argues, when one has no choice, that is, when one is obliged 'to combat dark desires'. A person possessed by such desires instinctively distrusts the ability of both themselves and others to obey the rules of the Apollonian social bond and so contain Dionysian urges. What is required is the added reassurance of an all-seeing reason endowed with the power to enforce obedience always and everywhere. Platonic thought thus marks the admission that one no longer trusts the power of one's conventions, customs, or habits to contain unsocial inclinations. Convention and selfhood have become irreparably disassociated from one another. Socratic reason is the attempt to correct this separation by wrapping the social environment and the body alike within the all-encompassing net of a conceptual ordering that imposes abstract rules so as to generate a fearful, schematic cage of light. The decaying self is encased and preserved in a bandage of method.

102. See Theodor Adorno and Max Horkheimer, *Dialectic of Enlightenment*, ed. Gunzelin Schmid Noerr, trans. Edmund Jephcott (Stanford, CA: Stanford University Press, 2002), 1–35. For Horkheimer and Adorno, the Socratic proclamation of a reason capable of providing an objective standard for determining human ends eviscerates itself in the modern era as the direct consequence of modernity's pursuit of rationality. This pursuit subjects the concept of objective reason itself to criticism and finds its claims wanting. What is left over after this is a purely methodological conception of reason, i.e. reason understood as something akin to a mere instrument. Such rationality is useful for facilitating ends but helpless with regard to assessing the desirability of any particular chosen ends. Morality, in consequence, gets rendered a matter of taste akin to personal preference. Modern reason, as Horkheimer memorably puts it, thereby exhibits 'a tendency to dissolve its own objective content' (*Eclipse of Reason* (New York: Continuum, 1974), 12–13).

103. Adorno and Horkheimer, *Dialectic of Enlightenment*, 20.

104. For doing the right things will, Socrates always tells us, make us happy.

105. Plato, *Republic*, 414bff. The lie in question is the naturalization of the belief that people are born with either gold, silver, or bronze in their souls. Those who are gold-souled are fit to rule; those who are silver-souled are fit to defend; and those who are bronze-souled are creatures driven by appetite who are fit to labour.

106. Plato, *Republic*, 540a.

107. Plato, *Republic*, 484c.

108. Plato, *Republic*, 517c.

109. One thinks here of A. J. Ayer whose faith in the methods of science precluded any critical reflection on its limits (*Language, Truth and Logic* (London: Victor Gollancz, 1967)).

110. See my discussion of this in the introduction.

111. This is, in effect, the position developed by Giorgio Agamben in *Homo Sacer*: *Sovereign Power and Bare Life*, trans. Daniel Heller-Roazen (Stanford, CA: Stanford University Press, 1998).

Part 2

FOUR CHAPTER-LENGTH SKIRMISHES WITH HYPERMODERN EYES

Chapter One

Five Short Pieces on Screens

Cell phones, but in every other respect medieval.[1]

Ikon

I, ikon,
accept your mass
e-reverence,

your graceless
attitude of prayer,
all arms and thumb-

nails, dumb iCloud
of witnesses. It's what
this pitying smile

was painted for.
I suffer you
to see me

through a glass
and through a glass,
darkly. Draw closer

now; cast off
those little tablets
of the law, your tiny

shields of Perseus;
look me in the eye;
come naked

as that bonelike self-
ie stick and I will
vex you face to face.

An image always shows a world. Worlds have their constituents, some of which they share in large part with other worlds, some less so. Some constituents stand out more than others. This does not mean that what stands out most is always what is most important. Here we find ourselves seeing a large wall, although perhaps this is not initially what draws our attention. The wall is mottled, but that aside it consists mostly of a climbing blank edged by a couple of neat looking borders that fade up into blur. It is almost as if it is leaning away from us. Glued to the wall's face is a female face in a frame (enclosed behind an invisible wall of glass). Seen in such a context, she is diminutive. She appears to look back out into the indistinct crowding throng before her. Or is she looking over them? Gazing through them, perhaps? She is posed, oblivious, iconic. She cannot return the gaze, yet such is the specularity of our vision that we really cannot avoid the feeling that we lookers are being 'looked into'. There hangs Mona Lisa. We probably know already 'where' we are. There are also those who congregate about her frame. They are the onlookers, a dissolved multitude, anonymous, a milling of temporary bodies caught in a slice of time before something that is more enduring than they. A solitary, sidelong person's glance aside, all heads appear to be turned toward the painting. We can imagine that they are active. Possibly they are absorbed too—though in what will be one question to ponder. We can see that some bear devices with glowing screens: apparatuses of capture. Mostly these visual prosthetics are held up in hands. One perches on a selfie-stick out of sight. This provokes imaginings of birds on branches or of the clawed mechanical hands that save those who need it the effort of reaching up for something. The machines are aimed at capturing something about the picture before them, about 'being there' alongside it. They are little packets for storing up presence. It has often enough been asked, Who is Mona Lisa? Historians of art can seek an answer by searching for the long since dead persons behind the daubed persona. Machines can scan the image's surface for hidden drafts and alterations underneath that give clues to the construction. She is iconic. As icon, though, she is no person at all. No 'who' corresponds to 'her'. 'She' is paint organized around a certain thematic of personhood, a bundle of layered hand-strokes of the artist now mottled with age. It can be asked, once again, Who is Mona Lisa? But more to the point here, should it not be asked, Who are they who seek her by seeking to capture her? And, since we, too, are looking on and thereby implicated in virtue of our gaze, this is also our question: Who are we?

SECTION 1: MEMORY-PROSTHESIS

Roam around a gallery today and you will see persons acting in concert in hypermodernizing environs which stretch far beyond the supposed immediacy of their bodily sensations. Today the people in the Musée du Louvre who ponder Mona Lisa stand before her already electronically connected—they fan out of themselves inexorably by way of the screened devices they hold. We (most of us) are 'they'. In this fanning-out they also participate in enclosing themselves within a structure of communication and observation that, because it now constitutes the world of the everyday in decisive ways, transcends them. They are, in this regard, simultaneously 'here' in the gallery but also at the same moment 'elsewhere', here and not here. In hyper-connectivity, the 'inside' and the 'outside' of the museum are conjoined as never before through the instantaneity of the ubiquitous communications device. Around the gallery people walk and pause and walk—a 'we' drawn to iconic artworks that have likely already been seen and contemplated (however casually), and so 'visited' without the need to have been there before, on computer screens.

Works are now for the most part encountered in this way far more than through images in paper media (newspapers, magazines, books) or in the flesh.[2] Their iconicity is now bound up with the 'we' of connectivity. They are seen in the 'here' and 'now' of the gallery as objects already mediated by this other form of viewing and the expectations it gives rise to. One sees this in the image of people gathered about Mona Lisa. Here is a veritable throng: an assembling of persons looking to bear witness to something greater than they. They are almost Dionysian. There are the gazes turned upward toward the icon. There is the icon itself. Then there is what marks the occasion, storing it away for future memory: the phone's camera 'through' which the icon becomes a transferable object of attention. 'We' are recording and witnessing something, and trusting this moment of witnessing not to the vicissitudes of fallible personal memory but to a device of recollection (stored away for later, for re-collection), to a mind-independent memory that promises to fix things objectively by translating them into images, that vouchsafes to freeze the 'as it was' and keep it fresh-packed for visual consumption later, elsewhere, by oneself and others, too.

Doubtless, the shared mechanical tendency to visual-prosthetic memorization has been around more or less since the mass production of the relatively easy-to-use portable film camera (perhaps first the Kodak then a host of others). But with the electrification of the image, the shrinking of the imaging-device and, above all, that device's ability not just to record but also to send its 'captures' elsewhere more or less there and then, the relation of memory to the mechanical-visual is intensified. It patterns itself into the movements of bodies—observer as much as observed. Intensification comes with ease of dissemination, simplicity of use. It is also the consequence of reconfiguration. In being rendered electronically communicable, the image itself is transformed into 'information' (that is, into code). It becomes a reservoir of memory as a result of the easy retention, storage, and transfer which comes with the electronic mobile device.

It is amazing to consider how far this tendency has spread. One checks one's bank account, does one's shopping, manages one's 'personal' life with a phone. One looks at a flickering screen not a piece of paper; one pays for more and more by way of it, or even falls in and out of love through it. Our relation with it is so pervasive that it has become casual, thoughtless, a reality that goes on in the shadows . . .

What happens to us now is foremost refracted though this flattened, shimmering, visualized form of memorization. It is the unseen regulator of the visible. The screen is the thing. Hypermodernizing life is electronically transferable visual events shown within the measured confines of a display which, even as it points outward beyond itself, encloses everything 'inside' (what one might call the private domain of subjectivity) within its own memorial boundaries.

Return to the museum in thought for a moment or two more, though. There is being there (bodily presence). Then there is pointing the lens and the box it points into: the visual event of capturing the memory of having been there, of signalling one's having once been in a particular place through the interpolation of the memento afforded by the device. One makes of oneself by way of this an interface between an electronic world, with its specific kinds of cables and bodies and accompanying energies, and the tactile presence of the icon on the wall. It is consequently the device and its screen which becomes pressing in the presence of the artwork, at least as much as the work 'itself'. What, then, is the icon? Is it the work or the device? Ikon or Nikon,[3] Mona Lisa or iPhone? An archaic religious tendency to worship the image erupts in hypermodern form, in the fetishism that combines image and device.

SECTION 2: (W)HO(L)LY SECULAR?

Museums can be fertile places in which to uncover signs of an emergent hypermodernizing religiosity. They are popular and rich storage bunkers containing traditional forms of the holy. We value (and, dare one say it? should value) what they hold. However, it is hard not to notice that museum collections are overwhelmingly organized according to the Enlightenment idea that an unbridgeable gap separates the modern rigours of observational scientific method, which they themselves exemplify, from the artistic forms of expression and archaic religiosities which they curate. In this regard, the museum's dominant principle of organization tends to be far from neutral. It presupposes the autonomy and organizational primacy of method (counting, testing, cataloguing, and so forth) and arranges its spaces and materials according to a sense of purpose which harmonizes readily with that practice of methodological rationality.

At the same time, what the museum preserves cannot, by its own lights, be entirely in accord with this organizing principle. By separating religion and aesthetics from science according to the dictates of science, religion and aesthetics are supposedly drawn into the objective sphere of rational secular power as objects of knowledge amenable to being viewed at a distance by the objectifying rational mind. The aim in this context is to master religion and aesthetic feeling by permeating their artefacts with reason in order to see what they then reveal. Yet this situation is no one-way street. As has been perceptively noted by Carol Duncan,[4] the very presupposition of a self-contained analytical reason actually allows for the unacknowledged introduction of quasi-religious and archaic elements into the construction of the purportedly secular-scientific public space of the museum. As is most evident in their buildings, modern museums remain monumental institutions whose enlightened worship of reason invokes the sacred sensibility of ancient temples and cathedrals. This sensibility serves to endow collections (which are more than random assemblages of parts) with a quasi-religious aura. The source of this aura is, for secular sensibility, held to reside in spheres of historic and aesthetic value. This, in effect, endorses the humanist sanctification of human doing. The secular endorsement of the religious aura, in turn, authorizes the secular knowledge which claims the right to organize and evaluate the holy historical-aesthetic relics through which the sacred aspect of human identity is communicated. The religious, one might say, is in this way sneaked back into profane museum space as a hazy symbolism which renders that space holy. The question is this: Is the secular itself not therefore always already permeated by the ancient sense of the holy it here proclaims only to curate from an objective distance?

The meeting and dissolving of apparently opposed worlds (secular-rational and religious-aesthetic) is illustrated by the masterly ancient Greek sculptural fragment the *Winged Victory of Samothrace*.[5] As a comment by Paul Virilio suggests, there is in this piece an intermingling which unnervingly closes the gap that 'progressive' (i.e. positivistic) Enlightenment thinking would like to insert between modernizing secular reason and archaic religious mythologizing. The statue presents a vision of 'technological vitalism' expressed by way of the aesthetics of 'the transport engine'.[6] The unification of the winged goddess and a ship (a mode of locomotion) expresses an ancient ambition of transcendence by way of the overwhelmingly powerful technical device. The statue hence symbolizes a thought-technology which promises to rise up, overlook, and conquer all that surrounds it with the all-embracing power of a detached and untouchable gaze. This is a military ambition which modern science inherits no less than Platonic metaphysics did in its day with its worship of a detached, specular, representational reason: the ambition which feeds off the notion that 'true' knowledge is revelation achieved by the objectifying gaze of rational-technical reflection.

What is at stake here does not concern specific purposes. It is a matter of the power which facilitates a certain mode of purposiveness, of practices of domination which facilitate manipulation. Through the old statue one senses already stirrings of the surging power of the machine's promise of a detached observational standpoint capable of flying above the environment, encompassing and so dominating all that falls within the frame of its surveilling.

We crowd about, pausing, bearing witness. Click. The camera and the aesthetics of a powerful visual transcendence which is its ancestor come together: one machine of capture frames the other on its screen and thereby affirms the common logic that binds them together. The 'purposes' may be different: the flying boat device seems to be striving to rise up in order to look down so as to celebrate the overpowering gaze of military conquest; in contrast, the people, cameras in hand, seem simply to be orientated toward freezing a moment in time for posterity. The relation to the gaze, however, testifies to a commonality of practice which binds statue and user of the device to one another. The practice of looking endows the sense of power through the distance that detachment affords. Both worship the visual in its capacity for detached distantiation. In the raising of hands, heads, and devices necessitated by the elevation of the statue, it is hard to ignore an unconscious mimicking of the act of supplication. At the same time, in the image above this distance is collapsed—for the image offers a sidelong glance at what it observes, separating itself along the very plane of separation it seeks to expose. In such moments we can bear technological witness, as believers, to the ancient truth of our faith in our own equipment and thereby become sceptics, non-believers.

The raised cameras of onlookers signify a moment of re-unification, the return to view of this archaic sensibility celebrated by greeting and prayers of thanks. Given that a 'society in which the sacred is entirely absent is impossible',[7] perhaps no one should be surprised to find it here, in the midst of the secular museum space. Is that who 'we' are? Does, through us, an ancient promissory note of unassailable power attainable by way of the vision of a religious-military transcendence reappear in the imaging device's eidetic seizures, its grasping at a preservation of the world in terms of the distancing promised by a single and overpowering on-looking gaze contained in a screen?

SECTION 3: WHO ARE WE?

There are always answers on offer. 'We Are What We Do', says a recent UK branding campaign for a bank, slickly adopting tones of morality for consumption by the observers of glowing electronic screens. Doing rather than simply observing? The campaign's webpage seeks to link seamlessly the worlds of the personal and the corporate—watcher and doer. It conjoins them through the commonality of a 'we' which is made evident enough when it is set firm before us in the characters of hyperspace script: http://personal.natwest.com/personal/we-are-what-we-do.html.[8] One can read such sequences like any text. The script signifies a command 'understood' by a receptive computer. The commanding pertains to something in common with a military order. The commands which guide computers also guide missiles and bear a functional family resemblance to the commands issued to obedient soldiers. The computer command which guides the weapon guides the advertisement also in its passage through the competitive terrain of the electronic global marketplace. It consists of a sequence of marks relating to pulses of energy compelling a computer to display a 'page' associated with those marks.

Where the screen is commanded, the viewer's gaze is supposed to follow. This has the virtue of revealing the falsity of any system of belief which begins with the purportedly obvious premise that always and anywhere 'I am necessarily *here*'. 'Here' is a notion rendered complex and multiple by the screen, which transports the viewer to other places precisely by keeping them where they are. In doing so it multiplies their points of view, sometimes to the point of confusion. 'Here' is always more than one place—but then, in truth, perhaps it always was.[9] In the case of the advertisement 'We Are What We Do', the commanding link transports the viewer to a space that greets them with a kind of moral invocation that is rather piously knowing in tone: 'We all have a duty to do the right thing'. The viewer is thereby subjected to another command (ethical) that unfurls from following the pathway initiated by the first (digital), this time in the shape of an 'ought' made all-inclusive by the 'we'. The webpage has endowed itself with authority in tones vaguely reminiscent of a preacher. It seems about to deliver a kind of sermon, although it is scrupulous in its secular avoidance of overtly religious connotations. 'We' (the unity of the representation of corporation and viewer, since this is an all-embracing 'we') are bound and so formed by a common acknowledgement that there is a duty to do such-and-such (i.e. the 'right thing', whatever that may be). There is, it follows, a kind of implied moral 'rightness' that trumps everything else. This being the case, there must necessarily be implicit common agreement about the duty to do what is 'right' even before the 'visit' to the page, which will simply confirm that 'we' (the bank) are like 'you' (the viewer-customer) in this regard. We all inhabit the same common moral space—that is why 'we' are held to be one and the same. The only thing left to do is to transfer the terms of the normative commitment associated with 'duty' into a contractual one since implicit in the encounter is the request that you do business with 'us' (the bank). That is what advertising is for, after all. That is why the advertiser, the creator of the corporate image, exists.

As needs must, the space of the page radiates out beyond its immediately illuminated confines. One cannot see it without being confronted at every turn with temptations of connectivity. Looking at the page one discovers that there is, it turns out, something akin to a secular sermon waiting to be connected with. Clicking on an adjacent link brings to life an accompanying film advertisement which has also done the rounds on television screens. It consists of a mantra of contrasted self-descriptions which play with and deploy the notion of the 'we' by using pairings of concepts. Who are we? It is happy to proffer answers: 'We are [. . .]'—and then a list, accompanied by sparing, baroque, musical arpeggios borrowed from J. S. Bach.[10] We are told that 'we' are (1) 'thoughtful'/'thoughtless' (children sharing an ice cream/a man (largely concealed but most likely a man) emptying an ashtray out of a moving car window); (2) 'kind'/'cruel' (a woman feeding a tiny puppy/people, again probably male, forcing a snapping hound into a dogfight); (3) 'creative'/'destructive' (a woman's back covered in intricate tattoos/a skinhead man with 'hate' tattooed on his knuckles); (4) 'brave'/'stupid' (archive footage of two young soldiers smiling straight at the camera, one giving a thumbs-up, as they march nervously off to war/middle-age male hooligans ransacking a bar); (5) 'heroes'/'villains' (a protesting suffragette accosted by police/a covered criminal being bundled from court into a police van—again, one suspects a male); (6) 'visionary'/'blind' (a sitting-room television showing the logo-typical pathos of Chaplin's 'little Tramp' hunkering down for the night and making the best of it/the predicament of a homeless woman coldly ignored and thus passively victimized by passers-by, their legs sweeping in front of her on a dimly lit street; (7) 'responsible and *responsible*' (a woman carrying her sleeping child toward a house front door/the crumbling ice caps of an irresponsibly warmed-up world).

One might pose the obvious question as to whether it is female viewers who are this advertisement's primary 'target audience'. Be that as it may, it is interesting to note the pairings which accompany the flickering imagery. The contrasting terms and images are themselves not always intrinsically opposed. There is, for instance, no necessary opposition between 'brave' and 'stupid' (the brave are usually contrasted with the cowardly, the stupid with the wise—and plenty of people, one suspects, have existed who have been both brave and stupid). There is no incompatibility between the accompanying images of young conscripts and the barbarism attributed here only to brawlers rather than the warfare the conscripts are smilingly marching into. There is no contradiction between a loving parent taking her child into a centrally heated house and crumbling ice caps, since the one scenario with its carbon-spewing heating system humming away in the background is a likely contributory cause of the other. Each of these pairs, however, turns on asserting a different attitude toward care, and it is this which unites them thematically in opposition. Chaplin's perennially silent, suffering tramp preparing for sleep, his head lowering on a reclining dog for a pillow, offers a sympathetic imagining of the plight of the dispossessed which contrasts with our everyday and casual, brutal, out-of-sight-out-of-mind attitude to the homeless. Care is the unity of intimacy, the lack of it the disunity of anonymity. Intimacy is proximity, the individual standing out from the crowd by reaching out in one way or another toward us to touch us with our own vision. The images are unified though the generation of an organized cinematic intensity of feeling through which this intimacy of touch is projected out to occupy the attentive space of the anonymous viewer and create the condition of 'we'. Apparently, we are all in this together.

The advertisement's concluding voice-over tries to clinch things: 'We are all what we have done [an image of age, an old man], and what we will do [a picture of the epitome of innocent youth, a tiny baby wrapped in a blanket]. So we all have a duty to do the right thing [an image of a bank employee]. We are no different [we are like you]. We [the bank, like you] are what we do'. A feeling of 'authenticity' is achieved by reaching out in an appeal to the anonymous viewer's hopefully uncritical sense of their own moral rectitude. How many of us, after all, ever doubt our own moral superiority with genuine seriousness? This complicity generates the advertisement's sense of 'we'.

'We' passes from cradle to grave . . . well, actually, from impending grave to cradle since the sequence is presented here baby last, presumably because although old age is the outcome of every life-story not tragically cut short, for advertising purposes the impending death of the old is best only implicitly acknowledged. From cradle to grave, from the beginning to the end of life, the corporate and connected world is presented as achieving a fusion with the individual's. 'We' are like 'you' and thereby encompass you.

The 'we' of the bank is now a gigantic machine-like 'person'. It is a kind of 'Artificiall [hu]Man',[11] a Leviathan composed of natural ones, of individuals like 'you'. Here, speaking like Hobbes and recalling Zygmunt Bauman's account of liquid modernity[12]—of hypermodernization—is a 'dissolved multitude'[13] willingly arranged, obedient in coordination with itself. This multitude is responsible, reliable, and predictable—doing the 'right thing'. To prove the latter is the case, when he or she has had enough of the film, the visitor to the webpage can click around the page and ponder examples offered via an interlaced patchwork of 'good works' and 'good business' that harmonizes the unity of the corporate with the messier ambiguities of daily life which it promises to help out with. Corporate identity (which is rapidly becoming the norm not merely in 'business circles', including the business of sport, but at all levels of education and politics, too) is in this way fused with presentations of the particularized yet paradigmatic experiences of a householder flooded-out by a storm or a mortgage-seeker and the virtuous good deeds and 'best practice' exemplified by diligent, individual staff members (the individual employee's face as standard-bearer of brand image, as a kind of facial limb of the corporate body). All of them (like us), to recall the advertisement's punch-line, are presumably what he or she has done and what he or she will do. This locks them together in a unity of commitment that transforms deed and image into *deed as image* to create a sense of 'we'. The ethical assertion (the 'we all' that is the subject of 'duty') is achieved as a coordinated performance of these visualizations facilitated by electronic links (by electronic commands).

The eye is enlisted. Where eye goes, 'I think' follows. In hypermodernity one cannot think the corporation without this visual commanding component and the obedience that unfurls from it in the shape of the need to construct an identity (a public-facing 'we' of some kind or another) that is explicitly seen and recognized in the digitally interconnected world. The html command moulds with person. That is presumably why, like every large company and subsidiary and all-too-numerous individuals the world over, the bank has Facebook and Twitter feeds, the former boasting over 370,000 devout or not so devout (dare one say, religious?) 'followers'.[14] What is it that makes for a follower if not an unthinking, answering to a command? But what is it that possesses the authority to command under such circumstances? Hypermodern space; same old obedience . . .

SECTION 4: SO, 'WHO ARE WE?'

One asks again. This is a question that feels as if it ought to reach deeper or higher than 'mere' advertising could. A certain depth of anxiety is in evidence when Samuel P. Huntington poses it.[15] It is difficult not to feel some kind of sympathy with his anxiety, either. Huntington regards contemporary American culture as being undercut by the emergence of various alternative articulations of identity, be it of the 'other-national', 'subnational', or 'transnational' variety. The situation envisaged builds on Huntington's earlier and well-known 'clash of civilizations' thesis,[16] which concerns the new world order that emerges in the aftermath of the collapse of communism in the late 1980s. This new world order is based on commitments to beliefs associated with particular forms of 'civilization', that is, on attitudes which affirm the specificity of cultural identity rather than the generality of supposedly universal ideals. Political struggle becomes recast as a contest of civilizations (Chinese/Asiatic versus Islamic versus Western European/North American).

For Huntington, the new world order initiates a crisis which threatens American society by undermining its traditions. This is a hypermodernizing crisis of liquefaction. What has commonly been deemed good is also replete with evil possibilities: 'Modernization, economic development, urbanization, and globalization'[17] engender a 'narrowing' of identity that undermines cultural differences and dissolves the barriers that function to ensure social cohesion. Distantiated electronic vision gives rise to distantiated relationships. People increasingly interact by way of long-distance communications technologies. These technologies cross with ease the distances which once ensured the internal continuity needed for maintaining cultural and political orthodoxy. The use of communications networks does not merely threaten conventional national-geographical and political boundaries. It also promotes forms of mutuality which function according to criteria that transcend these boundaries. Alternative ideas of belonging rise up to subvert now weakened traditions. With the rise of these alternative notions of belonging, the 'we' of 'We Americans' begins to dissolve. What we see on our computer screens, in short, changes our sense of spatial relatedness and with this comes a radical reframing of those whom we identify with most.

Huntington is expert in charting the curious reversal that this situation enacts. Some of the powerful new identities cultivated within secular hypermodernizing conditions appear to hark back to distinctly un-secular, un-modern sources. We live in times marked by an unsettling re-awakening of religiosity: 'The twenty-first century [. . .] is dawning as a century of religion [. . .] Violence between religious groups is proliferating around the world. People are increasingly concerned with the fate of geographically remote co-religionists'.[18] The truth of this contention appears to be made plain enough by the coordinated Islamic fundamentalist terror attacks on the United States on 11 September 2001. With these attacks, an iconic and ugly spectacle opens out on to a new, unpredictable age of religiosity. The moments captured on cameras of civilian planes being crashed by suicidal killers into the towers of the World Trade Center in New York symbolically mark the violent supplanting of the twentieth century's orientation toward 'ideological conflict' by 'a new era in which people identify themselves [sometimes with uncompromising aggressiveness] in terms of culture and religion'.[19] For Huntington, American society is thereby confronted with a situation in which it can no longer blithely rely upon the presupposition of its cultural supremacy through universal acceptance of its key ideals ('freedom', 'democracy', formal 'equality' before the law, progressiveness). Other cultural players will be playing a different game. America, he holds, must therefore actively assert its own identity in order to remain a successful competitor in a future 'multicivilizational' world that will be ever more resistant to the notion that Western beliefs are universal truths.[20]

The perceived rekindling and rise of the ancient fires of religious belief is Huntington's main interest in more ways than one. On his conception, one must fight fire with fire, which means one must confront religion by affirming religion. Huntington's strategy, in other words, is to replicate the logic that confronts him and turn it against it. He urges a decisive turn to the combination of Christian religious and nationalistic sensibility which he holds now, as in its past, is a strength which 'distinguishes America from most other Western societies'.[21] Nationalism is licensed here to seize faith and, in perfect harmony with Machiavelli's advice in the *Discourses*,[22] make it a tool of political power with which to ensure the integrity of social space. Religion, in other words, is put to use as a weapon of defence. The need for such a strategy is simply a sign of the times. It is almost natural: 'In a world in which religion shapes the allegiances, the alliances, and the antagonisms of people on every continent, it should not be surprising if Americans again turn to religion to find their national identity and their national purpose'.[23]

The observational tone here should not be confused with observation. What is at stake is not what Huntington sees but what he wants his readers to see, i.e. believe. It is a recommendation that asserts a view of culture and politics in line with the sovereign, exclusionary logic of the bunker mentality. The loss of boundaries as a consequence of electronic incursion is to be countered by a religiosity that sanctifies national-cultural borders, endowing them with a spiritual dimension to form a resistant, enclosing barrier of unswerving patriotic belief. The presupposition is that societies are unified cultural domains and as such survive and prosper only so long as they are closed off from external threats: only when a firm line of demarcation is asserted between 'inside' and 'outside' is their continuance assured. Since the loss of boundaries as a consequence of electronic incursion is no less dangerous than any loss of physically policed borders, boundaries of identity, too, need to be policed. The religiosity that is said to feed fanatical religion globally is in this way enlisted (as external threat) to sanctify American cultural identity, transforming that identity into a resistant theological fortress. Religion is, in this way, re-affirmed as a kind of military strategy.

Huntington's response, however, underplays the extent to which the crisis that he has identified is itself intimately linked to cultural forms closely associated with the historical dominance of the West. One needs to consider, above all, the contemporary unfolding of these cultural forms in relation to how information technology, dominant conceptions of knowledge, multinational corporations, and the globalized marketplace conjoin. It is this conjoining which gives rise to conditions which overrun and decode locales through the encouragement of ceaseless transparency of movement, calculative instrumentality, and endless flows of images, 'information', and opinion, giving rise to the phenomenon of what Huntington calls 'narrowing'. The overturning of 'tradition', as Huntington understands the term, is underwritten into the free-market civilization and social order he espouses. The 'narrowing' he fears has already happened by way of that order's cultivation of an environment which overspills borders by way of its ceaseless dissemination of images and signs which shrink the world's cultural and political geography in an enveloping knowing in pursuit of profit. Western hypermodernization, in short, plays its part not merely in fostering the overspill of techno-religious fundamentalism but in creating and propagating it.

How is the hypermodern implicated in fundamentalist sentiment? The often reproduced and horrific images of the planes crashing into the World Trade Center stand as compelling symbols of the over-spilling of borders with which fundamentalism is readily associated. It would be easy to conclude, as Huntington does, that by way of this spectacle some straightforwardly pre-modern meanings and purposes have been unleashed that are capable of unsettling contemporary experience, giving the lie to secularism and re-establishing religious faith as a condition of global political existence generally. Perhaps that is why he locks on to the potential such a spectacle affords for re-cultivating the Christian fundamentalist sentiment needed to 'rebuild' a sense of American identity. This, however, ignores the fact that the power of this 'religious' phenomenon draws on a sensibility that is made available only through the fluidity of the hypermodern communications networks that are not merely the passive vehicles through which a pre-given meaning is circulated. What is at stake in fundamentalist religiosity is not an unchanging 'meaning' or faith that springs from an archaic source located in divine will or deep human need. Meanings which today advertise themselves by calling on an autochthonous religious power have a magical and mysterious religious aura only because that aura is powerfully present and cultivated within contemporary technological culture too. A similar aura is detectable in the computer-manufactured 'magic' of contemporary fantasy films and violent video games. Intense hypermodern fundamentalist faiths of whatever kind partake of a primitive ethos bound up with technology—which technological entertainment feeds off. Technology facilitates a spectacle of illusion which generates feelings of a visionary Dionysian bodily intensity amenable to being cast in terms of religiosity. These feelings are no less hypermodern than 'religious', which means the religiosity in question is necessarily in some senses utterly profane. As has already been argued,[24] it is not purposive 'meanings' that endure but procedures in virtue of which such meanings are generated.

Technologically reproduced images of violence are instructive in this context. Such images do not simply record and reproduce violence. They also perform and thereby sanctify it by way of ritual. The persistence of the ritual is the precondition of the insertion of purposes into the scenario that the ritual opens up. Something primitive within the hypermodern shows itself in those moments when technology gets turned against human bodies in order to hurt. The visual 'representation' of the spectacle in its capacity as spectacle affords an instructive example. A hypermodernized spectacle of violence appears when an image is transmitted in an instant and reproduced continuously, rebounding the world over, permeating the environment and so becoming part of it. This speed of transmission and reproduction allows for an exposure of the myth that underpins the dominant contemporary notion of 'progress' that abides alongside faith in technological advancement. The myth is that technological refinement is an unquestioned good which helps distance humankind from the original primitive violence of its past.

The notion that technological advance is progress is placed in question when sophisticated image technologies facilitate the dissemination of spectacles of violence and so reveal their entwinement with them. It is this, not multiculturalism and the perceived threat of a 'multicivilizational' global dissolution, which poses a challenge to Western secularism. The challenge is mounted by elements within secular culture itself that can be turned against it. It is a challenge that reveals itself when secular social order, by endlessly reproducing images of violence on screens, sanctions the assault of technologies on bodies not merely by bullet and bomb but by way of the showing and circulating of the mass-communicated image. 'Showing' is never merely representing; showing is a practice which involves *doing*. Doing makes a world. The weaponization of the image deployed down the screen conduit is a key political ingredient in the world of contemporary culture. Culture, in this way, extends the violent potential of technology by reinforcing the notion that visuality, too, is necessarily a field of conflict. The electronically reproduced image is the contemporary weapon in virtue of which this field of conflict is opened up by exploiting the reproducible image's potential for achieving a visceral visual impact by provoking a bodily thrill.

Perhaps fear of and in technology *through* technology, not of religion, is therefore autochthonous, for the fear of violence is unleashed most effectively by technological means. Fear, after all, conditions the emergence of the false historical awareness evident in the feelings of religiosity conjured by popularized images of extreme fanaticism and terror that appear on screens the world over. This applies to Huntington himself when he is driven to replicate and so internalize fearful religiosity with his mirror-image representation of a redemptive Christian patriotism. Jürgen Habermas puts his finger on this powerfully enough when he notes that 'in spite of its religious language, fundamentalism is an exclusively modern phenomenon and, therefore, not only a problem of others'.[25] It is, in other words, Huntington's problem too. What occurs in the discordant contemporary entanglement of technology, religion, and fear is, as Habermas points out, not a clash of civilizations but a 'fatally speechless clash of worlds', a clash facilitated by the visceral power of the image to provoke fear. 'Faced with a globalization imposing itself via deregulated markets', Habermas continues, 'many of us hoped for a return to the political in a different form—not in the original Hobbesian form of the globalized security state [. . .] What we are left with, for the moment, is little more than the bleak hope for a cunning of reason—and for some self-reflection'.[26] Why, however, cleave to hope for reason so long as reason is content to be technology's willing and uncritical bedfellow? So long as technologically dominated rational cunning is complicit in the violence its technology creates and shows by way of its screens, it remains part of the problem, not a solution.

SECTION 5: DOWN IN THE MUD . . .

Foremost, we hypermoderns like to feel we reveal ourselves in the electronically generated public visibility facilitated by the communications device and its omnipresent screen. The common term for this is 'virtual reality'. Yet, for the most part, our understanding of what this 'reality' amounts to remains vague. Narratives of identity proliferate in high-speed bursts through what most people, following the dominant metaphor, think of vaguely as an electronic 'cyberspace' consisting of curiously floating, disembodied mediums such as 'storage clouds'. But these 'clouds', like the rest of the network of which they are a part, do not occupy any incorporeal realm of intellect or spirit. Clouds, like mists, are bodies. The world of the 'cloud', no less than the hidden passageways which make up contemporary city sewers, is a physical cultural structure of interconnected machines. Like all things cultural, the visibility contemporary culture here brings to the foreground tends to conceal a vulnerable materiality. Our vision makes us blind to the materiality of the visual. In spite of the seductive notion of a purely 'virtual' world, electronic visibility is facilitated by devices linked to one another by mundane pipes and cables, most of them sitting in mud on beds deep undersea.[27] They are vulnerable to being gnawed at by sharks. Cut all those cables, and the invisible condition which founds hypermodern visibility will suddenly, briefly, show itself as a jarring moment in which the vulnerability of the virtual 'everyday' is made manifest: the screen will reveal itself in its blankness for what it is.

NOTES

1. Jonathan Miller, UK Channel 4 correspondent, Channel 4 News, 25 May 2017. Miller was speaking of events he witnessed in the Indonesian province of Aceh: police publicly flogging two gay men who had been seized while together.

2. The intimacy of our knowledge of many works has for some time been a matter more to do with media-perception than with so-called immediate perception. One generally 'knows' the painting from the reproduction, or the symphony from the recording (few people ponder musical scores).

3. One should note here that many 'advanced', large-sensor cameras have also Wi-Fi connectivity allowing them to connect to other cameras, smartphones, and printers.

4. See Carol Duncan, *Civilizing Rituals: Inside Public Art Museums* (London: Routledge, 1995), 6ff.

5. Like the *Mona Lisa*, this too is held in the Louvre.

6. Paul Virilio, *Speed and Politics*, *An Essay on Dromology*, trans. Mark Polizzotti (Los Angeles: Semiotext(e), 2007), 68.

7. Gordon Lynch, *The Sacred in the Modern World: A Cultural Sociological Approach* (Oxford: Oxford University Press, 2012), 128.

8. This is NatWest Bank. Currently, the majority shareholder of this company is the government of the United Kingdom, and thus the 'we' of the British people. This advertising campaign was initiated in late 2016. Its presence on the Internet is marked by the html's binding of the 'personal' to the image of the 'we' that is being presented

to faithful and prospective customers alike: http://*personal*.natwest.com/*personal*/*we*-are-what-*we*-do.html (italics mine).

9. As Martin Heidegger notes in the early 1950s essay 'Building, Dwelling, Thinking', people

> persist through spaces by virtue of their stay among things and locales [. . .] We always go through spaces in such a way that we already sustain them by staying constantly with near and remote locales and things. When I go toward the door of the lecture hall, I am already there, and I could not go to it at all if I were not such that I am there. I am never here only, as this encapsulated body; rather, I am there, that is, I already pervade the space of the room, and only thus can I go through it' (trans. Douglas Hofstadter, in *Basic Writings*, ed. David Farrell Krell (London: Routledge, 1978), 359).

With hypermodernization, however, this condition is subjected to the power of the communications prosthesis, which initiates a collapse of locales by transforming the spaces one negotiates into 'information'.

10. The piece is Bach's Prelude and Fugue no. 1 in C Major, BWV 846 (1722), from *Das Wohltemperierte Klavier*. It is played very slowly on a piano rather than relatively rapidly on a harpsichord (as the even now relatively recent 'authentic' approach to performance would stipulate), the accompanying chords of an electronic synthesizer wafting along gently slightly beneath it.

11. Thomas Hobbes, *Leviathan*, ed. Richard Tuck (Cambridge: Cambridge University Press, 1994), 9.

12. See the introduction to this volume.

13. For an excellent discussion of the Hobbesian conception of the 'dissolved multitude', see Giorgio Agamben, *Stasis: Civil War as a Political Paradigm*, trans. Nicholas Heron (Edinburgh: Edinburgh University Press, 2015), 33ff.

14. See www.facebook.com/NatWest/ and https://twitter.com/search?q=%23natwest.

15. Samuel P. Huntington, *Who Are We? The Challenges to America's National Identity* (New York: Simon & Schuster, 2005).

16. See Samuel P. Huntington, *The Clash of Civilizations and the Remaking of the World Order* (London: Simon & Schuster, 1997).

17. Huntington, *Who Are We?* 13.

18. Huntington, *Who Are We?* 15.

19. Huntington, *Who Are We?* 340.

20. Huntington, *The Clash of Civilizations*, 18ff.

21. Huntington, *Who Are We?* 364. What happens to poor old secular Europe is not, presumably, Huntington's concern.

22. See Niccolò Machiavelli, *Discourses*, in *Machiavelli: Selected Political Writings*, trans. and ed. David Wootton (Indianapolis, IN: Hackett, 1994). Machiavelli notes that 'a good religion leads to good institutions, good institutions lead to good fortune, and good fortune ensures the success of everything one undertakes' (115). Consequently, 'those who destroy religions, undermine kingdoms and republics, are hostile to excellence, to literature, and to all the arts and crafts that are useful or honourable to mankind, these men are infamous and detestable' for 'anyone who reads the history of Rome with care will recognize how useful religion was when it came to

commanding armies, to inspiring the populace, to keeping men on the straight and narrow, and to making criminals ashamed of themselves' (110, 114). Machiavelli's point is simple. It both prefigures Huntington's and overturns part of its logic: religion has nationalistic value since it is a kind of social glue which binds the members of different orders to one another in a constructed commonality. If, however, the value of religion is essentially functional (as is revealed by its nationalistic appeal), then answering the question as to *which* religion one ought to advocate becomes a contingent matter. It seems odd, as Huntington urges, to foreground the national specificity of ideals whose power surely turns on their claim to a supra-national, transcendent, and eternal validity. The affirmation of national identity threatens to replicate the Hobbesian conception outlined by notorious legal theorist Carl Schmitt and characterized as epitomizing the political: the friend-enemy distinction. See Carl Schmitt, *The Concept of the Political*, trans. George Schwab (Chicago: University of Chicago Press, 1996).

23. Huntington, *Who Are We?* 366.

24. See the introduction to this volume.

25. Jürgen Habermas, 'Faith and Knowledge', in *The Future of Human Nature*, trans. Hella Beister and William Rehg (Oxford: Polity, 2003), 102.

26. Habermas, 'Faith and Knowledge', 103.

27. The telecommunications data company TeleGeography (www.telegeography.com/) offers a useful map of these cables. See www.submarinecablemap.com; and also Andrea Peterson, 'Everything You Need to Know About the Undersea Cables that Power Your Internet and Why They're at Risk of Breaking', *Independent*, 27 October 2015.

Chapter Two

17 Seconds

La Gare

Still the Beaux-Arts timepiece
reckons railway time.

Polished slabs and plinths
prompt langorous *dérives*
between the Baptist-child,
starved Ugolino and his bairns
and Mercié's *David*, slingshot-svelte –

but this is still *la gare*.
The shot insists on it,

pitching us along perspective's
standard gauge
to vision's buffer-stop
below Laloux's gilt clock.

All aboard, you coal-black boys
and glitching citizens.

What the passé Baptist's hailing
is the 4:08 from elsewhen,

hissing in.

Two seconds or seventeen? Just give two. A lot can happen in a couple of heart-beats. The entrance hallway on the ground floor of the Musée d'Orsay in Paris. There one can make a snapshot of two seconds' duration, resting a camera on a marble surface. Best to leave it to chance. Just the one snap, no second take. Four statues (not that one necessarily notices them first) in a zigzag relation that leads the eye into the distance. Nearest on the left, on a high plinth all but foot and leg concealed by the anonymity of being out-of-frame, is Alexandre Falguière's Le vainqueur au combat de coqs *(1864), then to the right Paul Dubois'* Saint Jean-Baptiste enfant *(1861) bearing his cross; behind and left again there is Jean-Baptiste Carpeaux's* Ugolin *(1863), famous as Dante's prisoner doomed to starve to death. Finally, right once more and further away still is* David, *by Antonin Mercié (1872), sheathing his sword, his foot on a severed Goliath head. Surrounding the exhibits is the luminescent aftermath of a passing of living figures, a flowing haze of bodies caught in the midst of moving from somewhere to somewhere. They are also rendered as statues or painterly smears. Some look on. Some are waiting. Some are resting. One, invisibly small until you magnify her, has her hand to her mouth. Another woman, half concealed by the large plinth on which Carpeaux's* Ugolin *rests, kneels in front of a child in a pram, yet another woman further to the left is dressed in a white shirt and speaking to a girl—again both tiny details. A woman and a man, one of them emerging from behind the near pillar topped by the slender statue leg of Falguière's statue, lean toward one another. He is wearing white-soled casual shoes; she wears sandals patterned with small holes cut out of them. They seem to be considering something, standing frozen deep in a conversation that appears like it cannot end. One can speculate pointlessly on the topic. The Carpeaux statue they appear to be looking toward? Or are they looking past it? A couple with backpacks reduced to a blur sweeps by on the near right, ghosts seemingly headed for the exit. All around are flows of movement, but they are not endowed by the frozen or smeared bodies alone. They spring from the tension engendered by the immovable surrounding structure, which closes in on nomadic mobility of people and things. There is the rising up of walls to greet ceilings. There is the concourse of marble and the columns and statues that jut out. There is the hidden closing-in of the institution itself. Movement is also engendered by the enclosure of the camera, a box that sucks in light and traps it with a movement of shutters. These elements keep hold of the nomadic forms that escape them only as afterglows of memory. Time passes; people pass. Near the centre a clock ticks. It is eight minutes past four in the afternoon. A lot happens in a couple of seconds, which means one might want at some point to stop counting.*

SECTION 1: SEVENTEEN SECONDS

Hypermodernity, it should always be remembered, has not happened. It is happening. It has been happening for quite some time and will probably continue to happen, for it is an ongoing phenomenon of cultural acceleration. It is not designated by a particular velocity which has been or can be reached. Hypermodernizing tendencies are made manifest in the kind of interest in velocity and its relation to time and aesthetic experience evident in a paper published in 2001 by Jeffrey K. Smith and Lisa K. Smith.[1] They take a broadly positivistic line of enquiry into the experience of seeing artworks in museums, offering a discussion which turns to the question of the relation between time and the museum. In accordance with what falls under the general rubric of 'empirical aesthetics',[2] their approach seeks to arrive at an understanding of visual aesthetic experience by converting it into the form of analysable data. Smith and Smith measure the time visitors spend looking at artworks in galleries. Their measuring of the durations of the viewings of 150 visitors to the Metropolitan Museum of Art before selected paintings reveals that people pause for a median time (i.e. what one might deem the amount a person might typically spend) of seventeen seconds before any particular work.[3] This, they say, generates a 'paradox'. Although people generally 'respond in glowing terms' to their visitor 'experience' and 'often find their visits to be extremely moving and positive', they nevertheless seldom linger long even before works 'that are generally considered to be masterpieces'.[4] Assuming visitors are not just being over-polite when filling in response forms (and doubtless there is a plausible case to be made for being plain deceitful with regard to such forms), intense experiences seem to be based upon encounters which are too brief to be anything more than superficial. One reason for this brevity of gazing, Smith and Smith hold, is because any 'museum is a palace of treasures that is limited only by time'. Time, it follows, is identified as the key factor governing the aesthetic experience of visiting a museum. They here make a point, reminiscent of John Dewey's contention, that the experience of 'art' via the museum cannot be of this or that isolated work.[5] The experience of art concerns 'the visit as a whole', for 'the collection', not the individual work, 'is the unit of analysis when spending time on art'. In pithier terms, when it comes to seeing art, 'the museum is the thing'[6] and we misunderstand how art is consumed if we do not appreciate this fact. The museum's time is an enclosure of visit and exhibit.

If the museum is the 'thing', however, this raises some interesting issues. For one thing, any particular artwork begins to appear to gain its power to offer the aesthetic satisfaction that provokes the eye primarily from being subsumed in the institutional structure which houses and surrounds it. The enclosure constitutes the gaze. The museum becomes the 'stimulus' not the artwork itself (assuming for the moment one can legitimately talk in terms of 'stimulus' and 'response'[7]). Individual works of art, one might say, come to rely in an essential way on the environment which supposedly serves them and derives the justification for its existence from them. The spaces which purportedly only 'frame' the works make them art. If this is the case, then what might be assumed by many visitors to be the mere means of engaging with art (the art museum) turns out to be providing the justifying purpose (the guiding meaning) that unconsciously underlies their visit since it is in virtue of the institution in which they are housed and to which the visitor is drawn that works come to be sanctified as 'art' and deemed worthy of being looked at with an aesthetically minded eye.[8]

One can approach what is at stake in such matters by bearing in mind Jean-François Lyotard's prediction that knowledge in contemporary technological societies has a tendency to be treated more and more like information. This means that times seem ripe for the rise to domination of those who only stand and count. 'Along with the hegemony of computers', Lyotard points out, 'comes a certain logic, and therefore a certain set of prescriptions, determining which statements are accepted as knowledge statements'.[9] This aspect of what Lyotard famously calls the postmodern condition[10] crystallizes a new pragmatics of knowledge and power. Knowledge ceases to be the possession of flesh and blood persons endowed with internalized and relatively independent memories and wills ('thinking subjects'). Instead, knowing becomes dependent upon a technology which transforms both knowledge and memory (since knowledge without the retention of memory is impossible) into code. The sites of knowledge and memory shift to a condition of externality determined by their relation to equipment. Knowing becomes conjoined with access to an abstract 'body' of online information. In turn, remembering increasingly comes to reside in the physicality of fingertips poised over real and 'virtual' keyboards which move in synchronization with eyes scanning ever more intently an environment revealed to them though programmable hardware. Knowing and remembering become the recollection of data that is ever more casually ready-to-hand and consequently also as easily forgotten as a casual assemblage of curios. With this, the 'constituted body of knowledge'[11] is destined to be dictated by technology and whatever powers are (or come to be) associated with it. This transformation suggests, amongst other things, that material historical conditions have a decisive influence on what has hitherto been thought of as the autonomous, 'mental' realms of knowledge, abstract thought, rationality, and taste. To 'know' something or to 'like' and 'appreciate' it, once subject to these conditions, means plainly to do so in virtue of a condition of electronic 'translation' which plays its part in determining what is to count as legitimate knowledge and taste and what not.

One can extrapolate from Lyotard's outline the implications this has for aesthetic experience and visuality in particular. With the forging of a powerful link between contemporary technology and vision, the latter becomes something ever more amenable to being accounted for in terms that accord with calculative computing techniques, the realm of numbers. The modern computer is in essence a machine organized around a screen that offers access to informational activity by way of the mesmerizing glow of its screen and whose power is amply testified to in contemporary life by its ubiquity. With this, vision, knowledge, and memory conjoin according to the requirements of the computer-demand that they become calculable information.

One might go further: the visual power of contemporary technology transforms knowledge 'itself' into something that is ever more visual in orientation, ever more a matter of the cultivated visceral sensation of the stimulated eye which it cultivates and shapes in accordance with the dominant form of awareness and self-awareness as 'information'.[12] Today, we like ever more to observe and count (which means both to produce and so become information, as well as to observe and count the information of 'others') because this satisfies a desire to 'know' the world we inhabit according to the dominant terms that govern it as a domain of display refracted to us by way of our devices and their visualizing screens. What Lyotard calls postmodernity is, in this regard, exemplary of hypermodernization: hypermodern acceleration collapses, liquefies, and saturates visual fields, colonizing and reinterpreting the practices which constitute them according to the dominant demand of speed. In this computerized condition of saturation, the visual is no longer separated from knowledge by the need to be mediated by abstract concepts entertained in a supposedly autonomous intellect. The visual becomes knowledge-of-information, i.e. knowledge itself, a knowing epitomized and validated by the presence of the luminescent symbol on the computer screen that conveys a sense of the energized cultural environment of connectivity that permeates it.

With the transformation of vision into information comes a corresponding possibility of re-articulating the aesthetic and the practices associated with it as data. Aesthetic experience becomes increasingly subject to the demand that, like information, it is easily reproducible and easily transferrable. To take this demand into the realm of the art museum, as Smith and Smith and others do, is to render its space paradoxical to a degree that exceeds the terms identified by them in relation to the notion of visitor 'experience'. For one thing, as they note, the museum remains 'the thing' that contains 'priceless' treasures whose uniqueness constitutes their value. Yet at the same time, as soon as one accedes to the requirement that these treasures be treated as parts of data sets and thereby reinterpreted and disseminated in that form, one transforms what is deemed priceless into its supposed equivalent as 'information'. The uniqueness of not just the work but the collection gets subsumed by the demand that it be translated into quantitative terms in order to be efficiently and effectively disseminated (e.g. the contemporary need for publicity that all institutions—governmental, educational, military—come to need in a world driven by communications fosters research which seeks to ensure that the museum, likewise, manages its assets as effectively as possible). The concept of 'uniqueness' comes to exhibit a tendency to be affirmed in accordance with the demand that what is valuable is no longer what is merely unique but also what can be easily communicated regarding its priceless qualities by technological means.

With this comes the hypermodernization of the art museum itself and the space of the gaze cultivated within it. Through online connectivity such space gets rendered both local and international in its immediate relations to those who engage with it. 'Local' and 'international' here become ever closer to being equivalent to 'information' too. The sense of place which the space of the institution traditionally designates flattens out beneath the hypermodernizing demand. Contemporary institutions today have been reconfigured as spaces of electronic and physical connectedness equipped with outward-looking 'exhibits' in the forms of websites and the varieties of interactive access these initiate. One can go in through the front door to bear witness to the uniqueness of the 'thing itself' (usually armed with one's phone). Or one can 'enter' the museum virtually by remaining safe 'at home' (wherever that is) within the comfortable parameters of the screen-sized space and dimensions of one's computer or mobile device.[13]

By transforming itself and thereby the works it contains into information to be delivered electronically, the contemporary museum increasingly emulates the logic of the global corporation (which is not to say that it is the 'same' as the global corporation) for it calls for our attention in the same way as the corporation calls for our money, which likewise become interchangeable forms:[14] both are subjected to the demand that they can be expended even when we are not 'there' in the immediate environs of the institution. In this regard, the approach of empirical aesthetics is apposite ('timely') precisely insofar as it accords readily enough with the dominant accelerating order of things, for the cultural space of the museum and the wider connected environment with which it conjoins all increasingly harmonize with the assertion that reality comes down to information, to forms of behaviour typified by velocities which stand in need of being measured in order to be invested with their proper, 'objective' meaning. There is, however, something enigmatic in this situation, something which slips through the grasp of the calculative gaze.

SECTION 2: ABORIGINAL MIS-ENCOUNTER

As a means of approaching this enigma, it is worth pondering further the exhibits of the contemporary art museum and gallery when it becomes subjected to the interrogative technical inclination of a hypermodernizing gaze focused on objectifying measurement. One can consider especially what *comes to* this kind of gaze *as it looks*, i.e. what it sees and what it does not see about its own evaluative presuppositions in the midst of its ambition to positivistic 'value neutrality'. It is a matter that can be approached quickly enough, and in a manner that builds on what has already been noted in relation to the work of Smith and Smith, by way of a volume published by Mihaly Csikszentmihalyi and Mick E. Robinson in 1990 called *The Art of Seeing*.[15] A brief consideration of their text begins to reveal what else might be at stake when one begins on the trajectory of pondering seeing as process and information that can be collected and counted.

Csikszentmihalyi and Robinson's approach is to theorize the aesthetics of visual encounters as examples of a more general category of 'flow experience'. They define a 'flow experience' as 'a heightened state of consciousness'. This state is held to be the common characteristic summing up what is felt whenever and wherever people are absorbed in 'sports, hobbies, challenging work, and social interactions'.[16] The idea is that although the perceptual stimuli which cause this form of experience might be different, 'the reaction itself, the structure of the reaction or state of consciousness, is the same'. On this view, there is no difference between being deeply absorbed in the activity of gambling or playing a game of football or encountering aesthetically a painting by Rembrandt, Van Gogh, or Jackson Pollock. The 'state of consciousness' (i.e. form of sensibility which determines conscious 'experience') is in each case a structural repetition of something which exists in common, for the content (that which the consciousness concerns itself with) is contingent.

The revealing of the structure of consciousness is facilitated by a statistically amenable format, that of the questionnaires which Csikszentmihalyi and Robinson use to conceptualize aesthetic experience. They invite respondents to express their feelings about the experience of seeing artworks in the gallery space in terms translatable into data. The deployment of observation and statistical analysis serves as a means of allowing the aesthetic to be articulated in broadly and self-consciously 'naturalistic' terms. Empirical findings are calculated, plotted, and graphed carefully enough.[17] This naturalistic approach, Csikszentmihalyi and Robinson tell the reader, accounts for aesthetic phenomena 'hedonistically' insofar as it allows any aesthetic experience to be defined as an experience which is felt to be 'good' by its perceiver.[18] Behind this lies the calculative notion that seeing aesthetically can be accounted for in terms of purposes ('goals'[19]) which can be substantively quantified. In other words, determinate meanings (realizable and consequently measurable purposes) are taken to lie at the heart of visual sensibility even though the 'what' that concerns any given purpose is coincidental.

Some of Csikszentmihalyi and Robinson's conclusions on the basis of this seem hardly earth shattering. They note, for instance, that cultural differences determine in large part the appreciation of art.[20] The reader ought, however, to be more forcefully struck by the terms chosen at one point to exemplify the condition of incommensurability that it is claimed separates different world views:

> It would be impossible for an Australian aborigine and a New York art critic to have similar reactions to an abstract painting by Jackson Pollock. The objective visual stimuli would be processed in entirely different ways by the two viewers [. . .] The two viewers would bring to the situation entirely different backgrounds of information. In fact, the aborigine might not have an experience at all in looking at the Pollock painting, let alone an aesthetic one.[21]

While the generalizing and culturally relativistic conclusion might be bland, the judgement rather offensive, and other of its aspects questionable even on its own terms (Csikszentmihalyi and Robinson seem not to have bothered to invite any indigenous Australians to fill in questionnaires about what they experience when confronted with a wall of Pollocks), what stands out is the contrast of standpoints that is constructed in order to make it. An aboriginal person, someone who is the direct descendent of a nomadic people harking back to the earliest of prehistoric eras, is contrasted with a Western metropolitan who is a visually literate sophisticate. 'Culture', here, comes down to the contemporary and distinctively Western notion of visual 'skill-sets'. Genuine cultural-aesthetic experience requires the requisite skills:

> without skills to recognize the possibilities contained in the artwork, the experience will remain shallow. Without sensitivity to the power concentrated in a well-made object, without interest in the feelings embodied in it, without curiosity about the existential context from which it sprang, even the most moving work will remain mute. Just one of these skills is enough to launch the experience, but in the absence of any, not much can be expected.[22]

The term 'skills' as it is employed here seems to be equivalent to 'tool'. In order to have an experience one must possess the conceptual tools which enable it. The possession of appropriate or inappropriate tools (instruments) is effectively held to divide one cultural community from another. It also functions to divide classes within communities.[23] Skills are conceptual equipment, cognitive technology. According to Csikszentmihalyi and Robinson's representation, being 'aboriginal' means being culturally illiterate in Western terms and therefore blind in multiple ways since the aboriginal person lacks the appropriate conceptual tools for the job. This, in turn, engenders a lack of the appropriate capacity for interest and, most strikingly, leads to the native Australian's being presented as devoid not merely of the ability for a specifically Western-style aesthetic experience but for any 'experience at all, let alone an aesthetic one' when confronted with the sight of a Western artwork. Such a person is at this moment thrust out of time as a consequence of being assumed to lack the skill-set (the bundle of conceptual equipment) required to make sense not merely of seeing this or that thing but of having an aesthetic experience *as such* as it is manifest in a typically and self-consciously 'modern' context (i.e. seeing the art of Jackson Pollock).

If the aesthetics of vision turns on the questionable attribution of a 'structure' of consciousness which denotes a condition of 'heightened awareness', then the indigenous Australian appears to stand accused of lacking not only this structure but the more generally elastic equipmental capacity that would supposedly allow him or her to adapt to different cultural contexts. The universality of the structure that is asserted gets determined by the instrumental context out of which the notion itself is articulated. Another way of putting this would be to say that cultural competences, in line with the tendency to think in terms of the primacy of information and technology as engendering authority, are instrumentalized through being subjected to the notion that what counts as 'real' experience is necessarily subordinate to the concept of 'information', which must be approached with the appropriate technical equipmental know-how. 'Skill' is shorthand for dealing with information (with experiential 'inputs', 'data', to speak in crude metaphorical terms). To recall Lyotard's point about knowledge, in the hypermodern condition what counts as worth knowing about aesthetic experience is what can be converted into data. Visuality has here become data understood as culturally heterogeneous bits of competence ('skills'), and this, as Lyotard notes, turns on a 'de facto legitimation' rather than on any autonomous objective or 'value-free' principles.

As befits an attitude dominated by skills (and positivism in general, for skills-talk is motivated by a commitment to positivistic attitudes), what comes to the fore is a methodological 'procedure' which functions by proposing what it presupposes to be the case as the criterion for establishing reality and then going out and looking for it in order to get the proof which confirms the notion of reality it has already committed to. The search is self-fulfilling and, in consequence, also excludes what does not accord with its demand for self-fulfilment. What counts as 'real' is determined in advance as what is amenable to technical recuperation and reproduction. A broadly 'technological' methodological attitude ends up in this way being used to produce its own justification. By reinforcing technology, one 'reinforces' the notion of reality associated with it, and one's chances of being just right increase accordingly. This is how legitimation by forms of power takes shape. Power is not only good performance but also effective verification and 'good' verdicts in virtue of which 'good' performance is codified reciprocally. It is self-legitimating—in the same way any system organized around the maximization of performance seems to be.[24]

If technology grew on tress and bloomed on cacti this might be less important. But it does not. Technology is made within a system of production whose relations conjoin interests and power. The point, however, is not merely the rather hackneyed (if nonetheless still relevant) insight that knowledge is power. *Information is power* and, insofar as it is amenable to being harmonized with information, *vision, too, is power*. The more the construction of an aesthetic discourse of visuality turns on information gleaned from empirical observation, the more it revolves around instruments ('skills') the possession or lack of which empowers or disempowers accordingly.

Implicit in Csikszentmihalyi and Robinson's instrumentalist judgement is a kind of Hegelian-spirited attribution of historical backwardness.[25] It is no good responding to this point by trying to claim that the situation would be the same were things reversed, that the Western 'expert' would likewise be left looking in a condition of blank, non-experiential embarrassment at an aboriginal creation. An exhaustive knowledge of the aboriginal person has already implicitly been claimed by Csikszentmihalyi and Robinson in their denial to them of even the possibility of *any* 'experience' when facing a Pollock painting. To claim to know what and when someone from another culture might or might not be able to see is to assert the power to situate them within the controlling ambit of one's own gaze, so subordinating them to it.

The primacy of instrumental forms of estimation is symptomatic of what, according to Wole Soyinka, can be called the Western obsession with conceptualization and categorization, a tendency which exemplifies 'a compartmentalizing habit of thought which periodically selects aspects of human emotion, phenomenal observations, metaphysical intuitions and even scientific deductions and turns them into separatist myths (or "truths") sustained by a proliferating super-structure of presentation idioms, analogies and analytical modes'.[26] Csikszentmihalyi and Robinson are so focused on assembling, counting, and analysing data that they fail to consider the cultural presuppositions that invisibly accompany the use of number. A sensibility of this order commands by setting in place a determinate space of the visual that is obedient to inflexible rules. Numbers and tables, which manifest the rule of calculation, are the tools and hence weapons which map this space not in order merely to represent it but to seize it. This is a sensibility which in this regard is, to recall Horkheimer and Adorno's summary of Enlightenment, at heart brutally 'totalitarian'.[27]

The notion of the incommensurable as it is deployed by Csikszentmihalyi and Robinson is exclusive. It enables an erasure and forgetting of the condition which constitutes cultures as conjoined realms (i.e. of the fact that cultures constitute worlds which do not open out on to environments in a way that hermetically seals them off from everything else that is 'different') even as it participates in that condition in order to notice the 'aboriginal' difference it then excludes. Culture is not one. But it does not follow from this that the multiplicity of human ways of living means that each form is closed-off from every other. Culture, like art, is enigmatic. Cultures essentially show themselves not just to the subjects whose identities emerge from them but also to other cultural beings (the one who is culturally Other) who can intimate something about them and on the basis of this learn from them. This showing is a constitutive condition of what culture is. It cannot be thought independent of this condition of multiplicity which necessarily includes those who are 'not like me' (which, unfortunately, does not preclude aggression toward them). This condition denies the option of attributing to any cultural standpoint whatsoever the brute and blank incomprehension that is best reserved for rocks and other things not endowed with understanding.

'Adventuring into Yucatan', Michael Oakeshott comments, 'I may perceive a "going-on" which, in perceiving it, I identify as the ruins of a Maya temple; that is, as a human artefact, an expression of intelligence'.[28] Of course, one may not comprehend the Maya temple beyond its announcing itself as a sign of intelligence which bespeaks a cultural presence, and such further leaps of understanding as one makes in order further to grasp that presence or what remains of it may reveal one's initial tentative judgement to be mistaken. One understands, after all, even when one does not do so appropriately—one understands even when one *mis*understands. But even in the midst of varieties of misunderstanding, the temple structure as the product of human doing is never entirely invisible to people without Mayan cultural competence (we will not say 'skills') regarding the religious, economic, or other aspects at play in it.[29] This is because what presents itself in the guise of the temple is not a domain of autonomous meaning that can only be detected when one is in possession of the appropriate calculative skill-set (the 'right' conceptual tools) but of intimations articulated in virtue of an archaeological residue of still-evident practices which invite speculation on meaning and purpose in a manner where no specific competence is needed. The temple's visibility (what one sees in seeing it) takes the form of announcing itself as an enigma of human action: it opens up a world which turns on ways of doing things that are not natural phenomena but spring from the body's articulation of its own relation to itself as culture.

Noticing ways of doing things is not, however, a matter of mere 'competence'. That is, it is not something one either has or does not have. It is the minimal condition of culture which is common to human understanding, although it dictates nothing beyond itself.[30] The temple's visibility takes the form of an invitation to understanding in its intimation of a world consisting of complexes of religious practices that inhere 'there' in the temple and in the echoes of observance that haunt it. The meanings one might attribute to it might be challenged, revised, refined, abandoned even, but the haunting residue of practice, the drama of its form which bespeaks intelligence, cannot be so long as the structure abides.

Culture always turns on the distinction between the products of the actions of intelligence (an intelligence which cannot simply be measured or quantified since it is that which enables measuring and quantifying in the first place and does so continuously) and those features of an environment that are not—between what shows itself in virtue of its being part of a world and what in the environment pertains to the worldlessness of stone. Bearing this in mind, Csikszentmihalyi and Robinson's analysis of 'objective visual stimuli' is more notable for its smattering of cultural imperialism than an 'objective' understanding of vision. Smith and Smith's urge to count time spent in front of artworks is not a million miles away from this in spirit. For Smith and Smith, Csikszentmihalyi and Robinson stand out only insofar as they 'argue that museums have the potential to cause visitors to give over their sense of time and place to the viewing of objects'.[31] None ponder the possibility that method is not value-free, that the investment in it entails damaging costs of its own. In the contemporary exercise of method lurks the potential to enshrine the remnants of an ancient universalism, the one that aligns sight and light with number and takes the organization of that combination to equate to the pure 'knowledge' of an unbiased vision. Here, in the midst of absorption in the empirical, is an old Socratic fetishism for the clarity and reassurance of finding a recipe for truth, a calculative mode of thought driven by the desire to make what is seen conceptually transparent by reducing it to the all-seeing pseudo-rational light which, in modernity and hypermodernization alike, is characterized by the turning ever more compulsively to the technologies of tables, graphs and statistics, and machinery to make the visible conform to the demands of the technically determined concept, the bit of 'information'.

An enigma of human action has already been spoken of in relation to the example of the Maya temple and culture more generally. There is also the aboriginal enigma of 'a direct, orally transmitted tradition which goes back for 40,000 years'[32] and which has endured in spite of attempts to exterminate it. This enigma, the enigma of survival in the face of overwhelming odds, the enigma of a movement which escapes the surveilling, predatory gaze of the methodical counter, suggests a conception of the aesthetics of vision which does not turn on a Socratism of royal method as calculation does. As Bernard Levin memorably notes, the aboriginal standpoint intimates a vision that entwines narratives of place to conjure an alternative pictorial imagining of the world: it turns on

> geographical legends, which are not simply prehistoric stories located in particular places, but stories made out of the places themselves; the land itself—valleys, peaks, rocks—is the language of the tale, so that if you stamp your foot on one of their sacred sites you are literally treading on a book which is at once their scripture, their myth and their anthropology. (It is also, they insist, their land).[33]

Grasped in these terms, space need not be regarded as a domain composed of calculated visual coordinates or points that are moved 'though' by individuated entities whose responses to 'objective' visual stimuli can be recorded, catalogued, and counted in order to arrive at a technical-statistical knowledge of the 'flow structure' which equates to aesthetic experience. What Levin hints at, in contrast to this, is something that consists of a world of practice in which the visible as such arises in tandem with the concrete experience of language as naming the world. To see the environment in this way is to see it not as something to be mastered by transforming the named into information, into 'representations', but rather as something which commands in virtue of being a living body of meaning made manifest as the visible environment. Such a conception suggests something that is 'non-technological' in the restricted sense that it does not transform the visual inexorably into data by making calculation the standard of value as such. With this in mind, one can turn, via Nietzsche, to the 'riddle' of art posed by Heidegger.

SECTION 3: SEEING THE ENIGMA

A whiff of the kind of positivistic empirical desire that motivates empirical aesthetics might be detected in a famous demand made in 1878 in *Human, All Too Human* by Nietzsche. What is needed, section 1 tells us, is 'a *chemistry* of the moral, religious and aesthetic conceptions and sensations'.[34] As is usual, however, Nietzsche's demand is more complex than it first appears to be. He also offers a hint as to how to get beyond 'chemistry', for his articulation of this requirement (like his naturalism generally) is enigmatic and ought to make anyone pause before he or she uncritically embraces the idea that methods and results are straightforward matters of application which concern a predefined domain of brute empirical data that can be applied straightforwardly to aesthetic sensibility. Making sense of naturalism involves strategies that are more complex than the mere taking up and putting to use of pre-given methods that juggle with numbers.

Method for Nietzsche himself, even in his most 'positivistic' period of writing, is always more a matter of metaphor than methodology (which is why *Human, All Too Human* has with some justification been called the first 'postmodern' text[35]). Take the core element of calculative method: number. Number, Nietzsche writes only a few pages after his initial, eccentric, and excessive naturalizing demand is stated, is based on error, i.e. on the belief 'that there are identical things (but in fact nothing is identical with anything else); at least that there are things (but there is no "thing"). The assumption of plurality always presupposes the existence of *something* that occurs more than once: but precisely here error already holds sway, here we are fabricating things, unities which do not exist'.[36] Number, in other words, is already bound up with a metaphorical substitution that is replete with unfounded ontological commitments[37] about synthesizing how a totality called 'reality' purportedly 'is'. Number is already an expression of culture's relation to the environment and not an independent encapsulation of the environment 'in itself'. Counting, in this regard, is at least as much an art as it is a strict and rigid 'method' for calculation, and the observation that arises with it is a cultural achievement. It is a way of coping with an environment by pragmatically opening out the possibility of a world of meaning onto that environment through a practice consisting of rules which foremost ensure a consistency of relations rather than offer us an absolute 'representation' of an objective 'reality'.[38] Pragmatic measures convince because they deal effectively with felt desires, pains, and the like and not because they are objectively 'true'.

What Nietzsche in effect suggests is that the domain of calculation in opening out human culture as a space consisting of measuring practices points beyond itself in such a way as to indicate that there is more to culture than counting. It may be that all of morality and much else of culture 'has its origin in the tremendous inner excitement which seized on primeval men when they discovered measure and measuring, scales and weighing'. Yet, with such 'conceptions they climbed into realms that are quite unmeasurable and unweighable but originally did not seem to be'.[39] Calculation is a condition of culture (and an invention) which properly comprehended intimates incalculable limits that exceed its formal bounds of knowing. Technological calculative practices, in other words, contain within themselves enigmatic possibilities which transcend them.

The nature of this enigma (or 'riddle'[40]) is explored by Heidegger when he identifies a tension at work between the instrumental calculation of technologically orientated thought and the non-instrumental comprehension that opens out of it in relation to the experience of art. A world, Heidegger argues, is not merely the empirical-sensory environment which technological manipulation seizes upon. It is a domain of significance secured in virtue of ways of doing things which make possible understandings, meanings, purposes, and the like.[41] The relation between understanding and technology is made manifest in relation to modes of visibility. Technology is equipment whose meaning resides in its propensity to be used up, and essential to equipment is the kind of familiarity that recedes into the invisibility which use generates. In simpler terms, technology does not show itself in the midst of the doing it facilitates. One does not feel one's 'good' shoes when enjoying a walk. Likewise, one sits at the keyboard of one's computer and types out one's thoughts by forgetting the typing and the keyboard, by losing awareness that the screen is a screen, forgetting that the chair on which one sits is supportive, etc. Keyboard, screen, chair, and all else that is equipmental disappear in the absorbed condition characteristic of putting-to-use, whose focus in any given instance on a given end obscures them. This invisibility has a tendency to shape the self in an especially telling way, naturalizing the conditions which epitomize it.

The problem for Heidegger with the modality of sight which emerges along with the familiarity characteristic of technology is that it betrays 'a peculiar tendency [. . .] which belongs to everydayness—the tendency towards [casual] "seeing"'. Such seeing is oriented not toward critical understanding but restricts itself to seeing '*just* in order to see'.[42] Seeing of this kind transforms what it witnesses into terms that are readily familiar. Artworks, in contrast, so long as they maintain their dimension as art, resist this everyday aspect of vision for they cannot be entirely colonized by the technologically orientated understanding which dominates the everydayness of contemporary life. The absorption characteristic of the experience of art is not the same for Heidegger as the kinds of absorption which might be said to typify other forms of activity. The artwork stands out and challenges us to think differently in virtue of its uselessness. As such, it raises the question of the value of a visuality which remains content with equating understanding with simply counting, recording, and sorting data and information that can be put to use.[43]

Art, in contrast to technology,[44] makes the familiar unfamiliar.[45] This unfamiliarity turns on the ability of the work to open up a world (a cultural space) and to make that world manifest by showing what is implicit in it. The work of art, in other words, engenders the possibility of critical understanding. The implication of this is that artworks do not simply 'represent' the social world from which they spring by conveying information about it. In their capacity as art they are not conduits for 'information'. Rather, artworks enclose the space characteristic of a social world in a manner that throws what constitutes that world into relief. Their enclosing of the environment, one might say, is not a defensive one—like that of a fortress or bunker—but, rather, a paradoxical entwinement in that it involves a closing inward that at the same time opens the cultural world out onto the externality of its environment. A structure like a temple, Heidegger comments, represents 'nothing'. It simply stands where it is. Yet in doing so the temple 'encloses the figure of the god'.[46] In this enclosing and concealing of space the structure allows the god's presence as cultural authority to be made manifest in understanding. Through this double gesture of hiding and showing the built structure gathers together its environment, endowing that environment with significance by articulating it through the practices which are implicit in the closure which marks out the temple's world and thereby making those practices explicit.[47] The environment is made manifest in its relation to a world of human concern, which is revealed as being what is genuinely at stake in the vision which alights and comes to life on the structure.[48]

Unlike the calculative making involved in technology, 'the temple-work [. . .] does not cause the material to disappear, but rather causes it to come forth for the very first time and to come into the open region of the work's world'.[49] The temple makes manifest the visuality characteristic of the culture through which the environment appears as an object of concrete concern. This condition abides 'as long as the work is a work', that is, as long as the structure relates to a concrete world of lived experience that has not yet slipped into the condition of archaeology requiring protection and conservation. It is the same with a text that engages in a self-reflexive contemplation of words or with an image (Heidegger's chosen image is Van Gogh's painting of a pair of shoes[50]) or with a piece of music. The essential feature of the work is that it 'set[s] up a world'[51] and reveals this setting-up by opening out an illuminated, variegated space of patterning, a 'clearing' in which things can appear and be engaged with from a non-instrumental perspective. Worlds, it follows, are events or happenings. They are not simply collections of empirical facts or processes to be observed and recorded by someone concealed in a discrete corner under the illusion that he or she is separated from them but occurrences which in being noticed demand to be understood and pondered by those who are drawn into their domain.

Art reveals the intrusion of the environment into understanding to be one of the essential conditions of cultural existence. The environment (the earth) is not passive, empirical 'material' waiting for willing technical humanity to seize and master it for its own purposes. The earth, rather, contradicts this willing urge. It can frustrate our demands and even devastate us (the earthquake, the tidal wave, the car crash) in a manner that places our instrumental ambitions in question. The artwork shows this tension in allegorical form by both refusing usefulness[52] and revealing a fundamental condition of conflict to be at work in the relation that pertains between culture and the environment. The world (culture) seeks to straddle the earth (the environment) and overrun it; the earth, however, always exhibits an unstoppable tendency to 'draw the world into itself and keep it there'.[53] Another way of putting this would be to say that no one (and no culture) can seek to overrun the entire earth with impunity since the earth continually looms up above us the more we seek to approach it with a view to containing it within the finite measure of our everyday existence.

A tension (a condition that Heidegger characterizes as a rift, rip, or tear which in opening and dividing endows a structure[54]) binds us to the environment we depend upon for life. Contemporary culture may be dominated by the tendency to figure the earth as a realm of boundless possibility that can be torn into and exploited by way of forms of measuring, calculation, and the like, but the environment resists. The earth is never pure usefulness. The earth outmeasures us because it exceeds measurement. And this is revealed by the consequences of our attempts to seize hold of it within the bounds of measure.[55] Thinking through this conflict between world and environment, for Heidegger, necessitates a turn to language. In poetry, especially, language displays the potential to turn to reflect upon itself as the abiding condition through which worlds of practices are opened out onto earthly environments.[56] In the poetic observance of linguistic practice which opens out in what has been insightfully designated the 'span of the poem',[57] the inherited continuance of our world is made manifest. Cultures, in other words, emerge in the naming of the environment. Naming, not calculating, articulates the space of a world as one in which people arise and can live. Art, in short, is a kind of enigmatic naming which makes manifest what culture conceals when it is taken up with the calculative curiosity that seeks to reduce the world, the earth, and persons alike to 'information'. To think art in these terms means to spurn the positivist tendency to reduce experience to categories of measurement. One must, it follows, begin to think through the limitations of this pseudo-naturalism in a concrete way. One can do so by thinking of beginning itself.

SECTION 4: OLD BEGINNINGS IN MODERNITY

What begins, Heidegger comments, is always at odds with what is familiar.[58] A beginning voices a tension within the everyday and the familiar. This tension can be generated by an artwork because art can erupt in the midst of familiarity's casually assumed, dominant mode of visibility, which is instrumental, disrupting and displacing it with a glorious uselessness. Familiarity simply repeats. Beginnings initiate. Familiarity smoothes over cracks; beginnings fracture. What begins draws on something other than that which merely abides in the here and now—it cannot be measured, counted, or theorized in advance of the particularity of its occurring. In a world in which use is a dominant mode of orientating visuality, the artwork stands out in its refusal of practicality. Its resistance to mere usefulness shows use from a standpoint which in defying use intimates more precisely just what use is.

This does not prevent artworks from being commodified any more than it can prevent thought from encountering the same fate. Even a poem can become fetishized as an object linked to a name (an original draft, a first edition). Likewise, the rich person's private collection of works can express the notion that art is a luxury whose uselessness is the mark of great wealth. At the same time, the luxury of uselessness can also stand as its own indictment since the attribution of uselessness is also the prerogative of the powerful which allows them to condemn the poor to being equivalent in measure to almost nothing, i.e. of lacking the 'social value' that is supposed to be the index for calculating the supposedly productive personhood that amasses private wealth. However, even uselessness which is commodified to a certain degree defies power. Its very existence as well as celebrating and securing privilege also threatens to turn any social order which glorifies the instrumental affirmation of value on its head: luxury, because it is lacking in use-value, is to be valued all the more highly than the usefulness which is as a rule valorized above all else. Usefulness has, in this regard, at least a doubled meaning which politicizes it and so opens up questions concerning instrumental value that can be debated and contested.

Familiarity is what we turn to in the pursuit of simply doing, or when we pause unreflectively in expectation that stuff will happen in the way it tends to in the normal course of things. Familiarity is also social hierarchy. The artwork's modality of showing stands out because it is neither a strict mirroring nor representation of this familiar reality (a picturing of the everyday world). It shows the world allegorically in a way that brings with it the suggestion of a lack—i.e. an intimation of the limits of familiar habits of 'representation' as such. A beginning, in contrast to familiarity, does not simply happen but is brought about within a structure of presuppositions which *show themselves as presuppositions*, that is, as moments of initiation which cannot simply be made use of in the manner in which things treated with the attitude of familiarity can. Beginnings, when they are grasped in this way, are paradoxical in that they seem to erupt as something 'older' than what is familiar, for the enduring hidden conditions of the everyday (the ways of doing things which accord with tradition or habit) abide within every beginning's questioning of them as a constituent of that questioning's possibility. However, they abide not as determinate 'meanings' but as hitherto unrealized possibilities of practice that can give rise to meanings which cannot be allotted a purpose or sense in advance of their coming. To see an artwork as a beginning is to recognize this estranged condition residing within it.

The artwork is a gathering of space which makes the cultural logic of enclosure manifest. Its uselessness shows not only the world of use (familiarity) refracted in allegorical form (i.e. a form which cannot be converted directly into use value) but also enclosure itself in its capacity of enclosing. This is not all an artwork can or does do or can be enlisted for, but this aspect constitutes the work's moments of critical possibility. In such moments the everyday familiarity of space is rendered unfamiliar in the midst of our daily comings and goings. In this way, art is one mode of opportunity for critical reflection on the world. These moments of reflection occur *between* things, i.e. between the constitutive features which make up an everyday world of concern. They thereby *show* these features to us in new ways, as a de-familiarization. In this regard, the work makes explicit what is implicit.

A time for a beginning always opens out from specificity. The present discussion began at eight minutes past four in the entrance hall of the Musée d'Orsay in Paris. One can return to it and follow the path of sight and thought begun there. It is a context that is at once an everyday one of familiarity and putting to use, and yet it can unfold beginnings which are located in the hypermodernizing environment but do not accord with its cultural logic.

SECTION 5: THE SPECIFICITY OF NON-PLACE

The discussion began in the Musée d'Orsay. One can return to it. There is uniqueness in the museum's space itself which is especially apposite with regard to the matter of beginnings. As the Musée d'Orsay guide tells you when you step inside,[59] it was the first art museum ever constructed (in the 1980s) within the shell of a disused industrial building. This older building was the Gare d'Orsay, a combined railway terminal and hotel which was constructed in the dying years of the nineteenth century to serve the Exposition Universelle of 1900's world celebration of industrial culture.[60] The Musée d'Orsay in this regard preserves and simultaneously transforms a quintessentially modern industrial space of ambitious speeding movement and use, erupting as it did from the energies of nineteenth-century industrial power into a contemporary art-space in a manner that brings with it acknowledgement of the value of the contrary ambition of exceeding such use.

The remnant in which the built structure of the museum is made testifies to the fate of varieties of speed in modernity and the discomfort this fate produces—redundancy, dereliction, disuse. Dereliction of this order tells us that industrialized acceleration is always bought at the cost of making spaces which are born with an aura of impermanence and outmodedness clinging to them. Like Little Father Time in Thomas Hardy's *Jude the Obscure* or Virilio's articulation of the bunker structures which pattern the French littoral, such spaces rise up prematurely aged. Accelerating modern temporality here exhibits its tendency to tilt headlong into the new with that ever greater eagerness that liquidizes itself in its own wake. Station space qua station is built to be made redundant—that is the price of 'Progress'. The gallery both reflects and resists this tendency by turning the temporal logic of modernization back upon itself, bringing out of its redundant industrial milieu a place for displaying the diversity of the modern arts organized in its collections of paintings, sculptures, photographs, design works, and architecture.[61] Heidegger, we have seen, writes of the emergence of culture as being marked by a rip which opens between world and earth. Derelict industrial space is one example of a rift of this kind in the form of the materiality of waste generated which outstrips its own usefulness. Here, in the museum, that torn space is reassembled and reintegrated. In this regard, it forms a paradigm of the ripped historicity of modernity whose gaze is obliged to look simultaneously back and forward in order to contemplate its own new yet damaged, battered, and worn(-out) self. Through the museum, the railway station (a room of exemplary industrialized movement) becomes an assembly point of the modern, an agglomeration which forms a momentary pause in its industrial flows rendered viable by reference to the authority of aesthetics. This assemblage is symptomatic of the process of hypermodernization and the anxieties it generates. The museum structure is constructed out of somewhere that for a while became a non-place in the midst of the city, somewhere derelict where nobody went—a place of uselessness. Considered as a railway station the building is quintessentially a piece of speeding modernity, yet it has already aged, too old to house the flows of industrial devices and commuters it once did. As a museum it preserves in the historicity of its structure this sense of non-place that disuse engenders as a kind of anxiety about what was and is lost.

Anxiety can provoke the desire to feel preserved and protected. As a form of enclosure, a museum is, after all, a kind of bunker—a space of protection that in being raised up simultaneously turns inward to create a variant of what has already been designated (at the end of part 1) Apollonian visual space. The promise of Apollonian space is of protection from outside threats founded on the condition of an inside composed of watchful 'publicness' which establishes civilized life by countering the dangers posed by the violence of Dionysian exteriority (the protection of the works). But it is in virtue of this space, also, that a Socratic conceptual enclosure of all-embracing, 'rational' light can arise in the guise of the seeing that would like to make of what is seen something akin to the category of 'behaviour' (the merely 'empirical' which is the object of methodologically inspired, aesthetic discourse). One must recall, however, that every Apollonian structure, however Socratized, at the same time contains the Dionysian within itself in the form of embodied, desiring creatures like us that are malleable but also resistant, seduced into being compliant but also potentially 'non-compliant' (to use a phrase resorted to by medical psychologists for patients who will not cooperate). The Dionysian is not simply the predictable 'behaviour' that is the beloved object cultivated by what Gilles Deleuze memorably calls 'control societies'.[62] Unleashed from Apollonian constraints by any gaze that would liquefy them, it is also lawlessness, the chaotic and unpredictable excess which contradicts even the 'lawful' causal structures which science projects into the environment in the form of 'nature' (physical, chemical, biological)—for the Dionysian operates in the discursive concatenation of such 'laws'. Around the artworks and the spaces opened up for their preservation and display, too, the struggle and tension between the Apollonian, Dionysian, and Socratic aspects of the visual plays itself out.

The space it opens out is permeated by an everydayness that expresses this tension, that is, an everydayness in which the unfamiliar which lurks within the familiar, which one can designate as *Dionysian*, also waits to be uncovered. Consider some of the people mentioned at the very outset of this chapter in relation to the photograph of the entrance hall of the Musée d'Orsay. They are practically invisible since they are enclosed by the image within the generality of all else that is occurring within it as small details. There is (from right to left) the woman with her hand to her mouth, the one who kneels before a child in a pram, and another who seems to be speaking with a young girl. One can pick them out.

Here are aspects of the world of everydayness. Situated in the midst of the posing of statues, we yawn, we rest, we wait, we look on, or we kneel and tend to the young. This seems all-too familiar, yet it is thrown into relief by the cultural space which de-familiarizes it. None of this is ever simply 'behaviour' any more than the diminished positivistic judgement which would make behaviour of it is. The presence of these people bespeaks a world in common, an intersecting of cultural experience (one does not have to be a European to have an 'experience' here). Speaking again in the language that emanates from Nietzsche's *The Birth of Tragedy*, we are Dionysian bodies taking shelter in an Apollonian space which opens out vision by endowing a sense of personhood that it nevertheless cannot entirely determine or control. Speaking like Heidegger, one might say that the intertwining of the placements of artworks with the subjects who move through museum space makes that space manifest as cultural space. But it does so in relation to them in a way that no amount of calculating and counting could.

In an image one can only detect this intertwinement as a kind of lack. As Svetlana Boym comments, the photograph is condemned to record movement as a lack, which shows up in a blur which 'makes the image at once nostalgic and poetic'.[63] In bodily movement, the space of the gallery reveals itself as an enclosure that opens up a world. It has already been said that moments of reflection occur *between* things, i.e. between the features which constitute an everyday world of concern. *Between them*, persons and works enact a world and thereby bring the environment to life—or, rather, bring the peculiar form of life called 'culture' to the environment. Thinking about them in this way marks out the possibility of 'beginning' in the sense intimated above and, alongside it, of a seeing which locates within the familiarity of place something unfamiliar, alien in its intimate relation to us. Only by presuming to 'look in' from the supposedly exterior (and, of course, superior) vantage point of a Socratic universalism, as if the world presented here were somehow separate from the gaze of an objective rational calculation which could distance itself from it, can this scenario be rendered in terms of the categories of 'stimulus' and 'response' which are amenable to being timed, measured, and objectified by the primitive Socratic urge to transcendence through calculative familiarization. In 'reality', which is not one place but always many (i.e. a kind of non-place for it can be seen from many perspectives and is in essence unfamiliar), we 'civilized' Dionysian bodies always 'communicate' more than just 'information' (data) that can be quantified, for culture when it permeates the Dionysian exceeds itself.

We are unfamiliar to ourselves, seen rightly. The Apollonian structure of the shelter at once frames and protects the people within it, but in doing so it makes them resist and stand out alongside the artworks that surround them in a manner that cannot be calculated or put to use any more than can the works themselves. Bearing this in mind, the calculative gaze of the empirical aesthetician might be well advised to reflect more critically upon itself than its casually predetermined objects.

What cannot be put to use seemingly wanders about. It perambulates. Perambulating, however, is not directionless and is not the same thing as being 'in transit'. A world in transit must always be articulated in terms of conduits through which people move on the way to other things, usually justified in terms that refer to substantive purposes. But the Dionysian elements as they perambulate, which have not been seduced into being at home in the cage of Apollonian-Socratic social satisfactions, are not headed anywhere in particular. They are nomadic, 'aboriginal' in inclination. They cannot be seen by being measured. One cannot *calculate* their gestures and translate them into statistically amenable meanings. Even a yawn, conceived nomadically, is not just behaviour.

We visitors are always already nomads even after we enter the 'civilizing' spaces of galleries. In the form of visitors and workers, we take our Apollonian-generated exteriority in with us and thereby inject these spaces with the life they would otherwise lack. Consider once more, then, Levin's pithy summary of the aboriginal notion of a vision that entwines narratives of place so as to construct a visuality that sees in the environment living elements of meaning at work. Such a vision refuses the archaic Apollonian distinction between 'inside' and 'outside' that constitutes the dominant mode of Occidental purpose—and hence, its colonial politics and vision. A way of seeing of this sort is not 'blind' to the domain of Apollonian enclosure and what can happen within it, as Csikszentmihalyi and Robinson's stance would suppose. It simply refuses its authority with a yawn. A nomadic narrative of this order transforms the landscape itself into the 'uselessness' of the artwork, making it stand out in its resistance to being reduced to mere stuff for use, to the status of yet another resource for protective measures to exploit in shoring up defences.

Within a nomadic narrative of this kind, trampling 'without a thought' on land would be no different from trampling on Mona Lisa's face and grinding it into the dust. Look out the gallery window. Turn the world the 'right way' up. It is raining. Ask yourself which way the rain falls. Down below, people traverse the urban space of enclosure which makes up the gallery's 'exteriority', impervious to the accusatory nature of their footfalls on the damp, resonant concrete surface . . .

NOTES

1. Jeffrey K. Smith and Lisa K. Smith, 'Spending Time on Art', *Empirical Studies of the Arts* 19, no. 2 (2001): 229–36.

2. The journal *Empirical Studies of the Arts* is the research publication of the International Association of Empirical Aesthetics which is a prominent exponent of this methodological-scientific approach. Empirical aesthetics seek to combine aspects of behavioural psychology, sociology, and varieties of evolutionary theory. Its roots extend back to the nineteenth century. It 'uses scientific methods to investigate aesthetic experience' (www.science-of-aesthetics.org/), although primarily 'it is most widely understood as a research domain within the field of psychology' (Fabian Greb, Paul Elvers, and Timo Fischer, 'Trends in Empirical Aesthetics: A Review of the Journal *Empirical Studies of the Arts* from 1983–2014', *Empirical Studies of the Arts* 35, no. 1:18). For an overview, see Colin Martindale, 'Recent Trends in the Psychological Study of Aesthetics, Creativity and the Arts', *Empirical Studies of the Arts* 25, no. 2 (2007): 121–41. For a fairly typical example of the empirical approach as it is applied to sight, see Andreja Bubić, Ana Sušac, and Marijan Palmović, 'Observing Individuals Viewing Art: The Effects of Titles on Viewers' Eye-Movement Profiles', *Empirical Studies of the Arts* 35, no. 2 (2017): 194–213, which offers an 'eye-movement data analysis' of responses to some Kandinsky paintings. See also Annukka K. Lindell, 'No Cheek Bias: Orientation in Van Gogh's Portraits and Self-Portraits', *Empirical Studies of the Arts* 35, no. 2 (2017): 127–38, which seeks to address the 'lack of a left- or a right-cheek bias' in Van Gogh's self-portraiture. Lindell has also investigated motor biases in selfies ('Motor Biases Do Not Influence Posing Orientation in Selfies', *Laterality* 22 (2017): 49–59). For another overview of empirical aesthetics, see *i-Perception* 3 (2012).

3. Smith and Smith, 'Spending Time on Art', 231. The mean time was 27.2 seconds, the mode 10 seconds.

4. Smith and Smith, 'Spending Time on Art', 230.

5. See John Dewey, 'Having an Experience', in *Art as Experience* (New York: Perigree, 2005): 'There is that meal in a Paris restaurant of which one says "that *was* an experience". It stands out as an enduring memorial of what food may be' (37). Aesthetic experience, likewise, stands out from the everyday, mundane realm of mere perceptions since 'the factors which determine anything which can be called *an* experience are lifted high above the threshold of perception and are made manifest for their own sake' (59).

6. Smith and Smith, 'Spending Time on Art', 234, 235. They also make the point that many visitors make more than one visit, which means that the cumulative experience of going adds up to more than a handful of seconds.

7. One cannot. See John Dewey, 'The Unit of Behavior (The Reflex Arc Concept in Psychology)', in *Pragmatism: The Classic Writings*, ed. H. S. Thayer (Minneapolis: Hackett, 1982), 262–74. As Dewey argues persuasively, 'the sensation and movement as stimulus and response respectively is not a distinction which can be regarded as descriptive of anything which holds of psychical events or psychical existences as such' (273). The distinction 'is one of interpretation' and presents a 'fragmented psychology' which ignores the manner in which we, no less than chickens, relate to our environment, namely, by way of 'a continuously ordered sequence of acts' (270).

8. These frames form an environment composed of social groupings, critics, museums, galleries, and so on, organized in a manner reminiscent of Arthur C. Danto's conception of the 'artworld' ('The Artworld', *Journal of Philosophy* 61, no. 19 (1964): 571–84). The artworld is an interpretative community (consisting of artists and critics, museums of art, art galleries, their owners and curators, etc.). What makes a particular work 'art' turns on its 'artistic enfranchisement' (573) in virtue of the embracing by the interpretative community of a new theory of what art amounts to.

9. Jean-François Lyotard, *The Postmodern Condition: A Report on Knowledge*, trans. Geoff Bennington and Brian Massumi (Manchester: Manchester University Press, 1984), 4.

10. Lyotard also refers to this as the condition of 'modern paganism'. See Jean-François Lyotard and Jean-Loup Thébaud, *Just Gaming*, trans. Wlad Godzich (Manchester: Manchester University Press, 1985), 15.

11. Lyotard, *The Postmodern Condition*, 4.

12. Witness, in this regard, the ever greater reliance in many aspects of life (business presentations or school and university teaching or social media) on computer 'visual aids' (e.g. 'PowerPoint', Facebook, Twitter) which seize the attention by seizing the eye of the beholder, translating knowledge and biography alike into chunks of light-information that can be fitted into the restricted space of the computer slide or screen. This might be well described as a fetish linked to the marketing of technological goods which discovers a rich seam of sales in the consumerization of education and other fields. Few currently tend to question the effectiveness of such technologies in spite of their obvious limitations evident to anyone who has sat in front of a slide 'presentation'. The PowerPoint slide is, after all, prone to become a crutch to be read out by the presenter, or worse, a distraction to the listener trapped in the dimmed classroom or lecture theatre, for eye and ear do not under such circumstances necessarily work in harmony as anyone who has read repeatedly over a projected text while trying to listen to someone speak is likely to know.

13. The websites of the British Museum in London, the Louvre or the Musée d'Orsay in Paris, and the Metropolitan Museum of Art in New York all reveal a powerful online presence. 'Virtual tours' are usually possible, allowing the viewer on a computer far away to 'see' the museum space by way of the luminescent screen. A good example is the Metropolitan Museum of Art's 360° Project (www.metmuseum.org/art/online-features/met-360-project).

14. See Claudio Celis Bueno, *The Attention Economy: Labour, Time and Power in Cognitive Capitalism* (London: Rowman & Littlefield, 2017). This matter has also been addressed insightfully by Jonathan Crary in *Suspensions of Perception* (Cambridge, MA: MIT Press, 2001), especially chapter 1. Crary is interested, amongst other things, in the manner in which modern culture creates the 'attentive observer' by generating distraction, which becomes a 'constituent element' in the creation of 'attentiveness in human subjects' (49).

15. See Mihaly Csikszentmihalyi and Mick E. Robinson, *The Art of Seeing: An Interpretation of the Aesthetic Encounter* (Malibu, CA: J. Paul Getty Museum and the Getty Center for Education in the Arts, 1990).

16. Csikszentmihalyi and Robinson, *The Art of Seeing*, 29.

17. Csikszentmihalyi and Robinson, *The Art of Seeing*, 97ff.

18. Csikszentmihalyi and Robinson, *The Art of Seeing*, 30. This instrumentalizes aesthetic experience from the outset since what is pleasurable constitutes the goal of the experience.

19. Csikszentmihalyi and Robinson, *The Art of Seeing*, 101: 'Item B15, "I have a rather clear idea of what to do when approaching a work of art", was the fourth most strongly endorsed item, with a mean of 3.6, closer to "often" than to "sometimes true". Directors endorsed this statement most strongly, assistant curators the least'. This, Csikszentmihalyi and Robinson note, is completely contradicted by another result where the same score was attained by another item 'measuring goals', namely, 'In approaching a work of art I never set some goal or objective I wish to achieve through the experience'. Interestingly, they account for this by arguing that it is the consequence of 'a faulty formulation of the item rather than to inconsistency in theory or the responses' (102), i.e. by cleaving to their faith in theory and method rather than questioning the validity of engaging with art in these terms at all.

20. They also tell us that amongst museum staff, two different 'modes of communication with an era or culture were distinguishable: one emphasized the differences between the past and the present, while the other emphasized the continuities' (83).

21. Csikszentmihalyi and Robinson, *The Art of Seeing*, 37.

22. Csikszentmihalyi and Robinson, *The Art of Seeing*, 266.

23. A point made by John Berger in his discussion of the notions of the 'primitive' and the 'professional'. For Berger, an art which looks 'primitive' is far from necessarily being so (*About Looking* (London: Bloomsbury, 2005), 71ff).

24. Lyotard, *The Postmodern Condition*, 47.

25. Hegel, Ngũgĩ wa Thiong'o argues, is an exemplary member of the 'tradition of great racists' who look upon other cultures and see in them something 'unhistorical', backward and 'undeveloped' (*Writers and Politics* (London: Heinemann, 1990), 14–20). What is being referred to here is the worst of Hegel rather that the best—assuming one can disentangle the one from the other.

26. Wole Soyinka, *Myth, Literature and the African World* (Cambridge: Cambridge University Press, 1979), 37.

27. Max Horkheimer and Theodor W. Adorno, *Dialectic of Enlightenment*, ed. Gunzelin Schmid Noerr, trans. Edmund Jephcott (Stanford, CA: Stanford University Press, 2002), 18.

28. Michael Oakeshott, *On Human Conduct* (Oxford: Clarendon Press, 1991), 13–14.

29. For what must count as the definitive account of, amongst other things, the economic and religious aspects of ancient Mayan society, see Robert J. Sharer and Loa P. Traxler, *The Ancient Maya* (Palo Alto, CA: Stanford University Press, 2005), 11:631ff, 13:719ff.

30. To be endowed with understanding means to be able to engage with one's environment in virtue of a world of ways of doing things (see part 1 of this volume). This is not some property a person can have or not have. It is intrinsic to selfhood although it in no way dictates the nature or character of the self, which is a matter that comes down to the specificities of time and place.

31. Smith and Smith, 'Spending Time on Art', 230.

32. Bernard Levin, *Conducted Tour* (London: Sceptre, 1988), 28.

33. Levin, *Conducted Tour*, 28.

34. Friedrich Nietzsche, *Human, All Too Human*, trans. R. J. Hollingdale (Cambridge: Cambridge University Press, 1986), sec. 1.

35. See Gianni Vattimo, 'Nihilism and the Postmodern in Philosophy', in *The End of Modernity*, trans. Jon R. Snyder (Cambridge: Polity, 1988), 164–81. Vattimo seeks to characterize the postmodern by turning to Heidegger's notion of *Verwindung*. As opposed to *Überwindung* (a critical overcoming of contradiction through recourse to reason), or a Kantian *Verbindung*, which seeks to establish a priori modes of the combination of concepts as a means of grounding transcendental analysis in universal principles of reason, Vattimo proposes an approach which performs a 'twisting' of meaning in order to make room for a relativized perspective which has a critical edge but disdains Enlightenment pretensions to objectivity. The 'post' in postmodern can thereby be accounted for in a manner that need not presuppose the possibility of transcendental critique. Nietzsche, he argues, fits into this scenario as the first thinker to write in such terms. 'It could', with this in mind, 'legitimately be argued that philosophical post-modernity is born with Nietzsche's work' (164).

36. Nietzsche, *Human, All Too Human*, sec. 19.

37. Unfounded, that is, if you want to say how things 'really' are.

38. Needless to say, what one pursues within the practice is not regulated in advance by the meanings that initially accrue to it.

39. Friedrich Nietzsche, *The Wanderer and His Shadow*, volume 3 of *Human, All Too Human*, op. cit., sec. 21.

40. Martin Heidegger, 'The Origin of the Work of Art', in *Basic Writings*, ed. David Farrell Krell (London: Routledge, 1978), 204.

41. See the introduction to this volume. As Heidegger comments, things in the environment do not present themselves directly to our senses in order to generate abstract ideas: 'We never really first perceive a throng of sensations, e.g., tones and noises [. . .] rather we hear the storm whistling in the chimney; we hear the three-motored plane', etc. Bare sounds and colours only ever strike us when we attend to them 'abstractly', and we can do so only because we have already engaged with them concretely ('The Origin of the Work of Art', 151–52).

42. Martin Heidegger, *Being and Time*, trans. John Macquarrie and Edward Robinson (Oxford: Blackwell, 2005), 214, 216. As Martin Jay notes, this distinction lies at the heart of Heidegger's attack on the 'occularcentrism' of both Platonism and its European inheritance (*Downcast Eyes: The Denigration of Vision in Twentieth-Century French Thought* (Berkeley: University of California Press, 1994), 270ff).

43. The artworld exemplifies the latter no less than the technically inspired thought of modern scientific method which is content to remain at the level of method. At the moment when an artwork has become familiar and is no longer suggestive of something out of place in the world, 'the art business has begun' (Heidegger, 'The Origin of the Work of Art', 193).

44. In this regard, Heidegger cleaves to the notion that technicity does not permeate the possibilities of human experience in a manner that dominates it entirely. This

notion is elaborated by the later Heidegger in 'The Question Concerning Technology' (*Basic Writings*, op. cit., 311–41).

45. Heidegger, 'The Origin of the Work of Art', 191: the less connected we feel to the artwork 'the more simply does the thrust come into the open that such a work *is*'.

46. Heidegger, 'The Origin of the Work of Art', 167.

47. The ground on which the structure rests rises up to support it; the inclement weather it endures is announced in relation to walls and roof. The environment (the earth) stands out and comes into view precisely because of the building.

48. Speaking in terms that relate to chapter 1, section 5 of this volume, one might say that the concreteness of the structure (a body) provokes the incorporeal flow of senses across its surfaces.

49. Heidegger, 'The Origin of the Work of Art', 171.

50. For a by turns illuminating, playful, entertaining, and provocative discussion of Heidegger and Van Gogh's shoes, see Jacques Derrida, *The Truth in Painting*, trans. Geoff Bennington and Ian McLeod (Chicago: University of Chicago Press, 1987), 255ff.

51. Heidegger, 'The Origin of the Work of Art', 169–70.

52. Heidegger, 'The Origin of the Work of Art', 145.

53. Heidegger, 'The Origin of the Work of Art', 174.

54. Heidegger, 'The Origin of the Work of Art', 188.

55. Technology when it is used up, for example, leaves an earthly residue in the form of useless or even noxious and dangerous 'stuff' such as the toxic waste that emanates from the tearing of matter which fuels the nuclear reactor or the polythene bags and containers whose constituents are prised from the earth only to be discarded to clog the oceans.

56. Heidegger, 'The Origin of the Work of Art', 197ff.

57. The phrase is the poet Bryan Martin Davies's ('Rhychwant y Gerdd', in *Cerddi Bryan Martin Davies* (Llandybïe: Cyhoeddiadau Barddas, 2003), 112). The poem opens out a space of possibility which is populated by words which make themselves manifest not only as words but in their painful testimony to the tortured, embodied, earthly conditions from which they erupt.

58. Heidegger, 'The Origin of the Work of Art', 201.

59. For an overview of the Musée d'Orsay's history, see www.musee-orsay.fr/en/collections/from-station-to-the-renovated-musee-dorsay.html.

60. For some discussion of the 1900 Exposition Universelle, see Robert Wilson, *Great Exhibitions: The World Fairs 1851–1937* (London: Studio Vista, 1977). The event attracted in excess of 50 million visitors.

61. The museum is devoted to modernity—its collections largely straddle the hundred years up to 1914, the first year of Eric Hobsbawm's 'Short Twentieth Century' (see *Age of Extremes: The Short Twentieth Century* (London: Abacus, 1995)).

62. See Gilles Deleuze, 'Postscript on Control Societies', in *Negotiations, 1972–1990*, trans. Martin Joughin (New York: Colombia University Press, 1995), 177–82.

63. Svetlana Boym, *The Future of Nostalgia* (New York: Basic Books, 2001), 21.

Chapter Three

~~Pixel Auto-biography~~ Pixel Auto-mythology

Digital Sin

This is the Tree of the Knowledge
of Pixel and Parallax, Filter

and Memory Card. He clocked it
on the road to Wick – black

talon of a tree. It hailed him
from the innocence

of analogue to digital's free will,
whispering *Take me, make me look*

as if the blasting god has just
passed through the Garden.

And he listened, primed
the camera for apocalypse

and clicked . . . lost paradise
in the knee-jerk shutter's fall.

What price memory? One might think, nothing. Memories, after all, seem sometimes to grow freely like weeds—one cannot avoid them mounting up. But also we cultivate memory to make our own individual and shared biographies. In cultivating we mythologize and reach back to pasts broader than our own. Consider the 'tree of life', a symbol that seems to be universally shared by cultures and religions—ancient Egyptian, Mayan, Jewish, Christian, Islamic. It is also extant in secularized form.[1] *One can gesture, by way of a contemporary example, to the exquisite fired-clay tree creations of Mexican folk artists which illustrate and celebrate the teachings of Christian faith. The trees are large criss-cross structures on which are mounted painted figures and shapes representing biblical stories: the creation of Adam and Eve, Christ's birth and death, and so on. Each tree is a reminder of Christianity's introduction from Europe into the Americas. Each is a call to renew religious commitment, its making itself an act of praise. The praise is richly creative, bursting out in an imaginative riot of colour and form. In its living tradition, memory and myth coincide.*

Here, however, celebrated by the poem, is what looks like a tree of death. Its trunk and branches seem like close kin to bleached bone. It juts out of earth, natural but somehow offensive in its demise for it is dead but seemingly refuses to obligingly vanish from view. The landscape is no cheerier: a monochromatic space of contrasts of light, dark, and half-tones, a clouded sky overhead and a bunched cluster of woodland in the background. The whole might appeal to the eye in its stubborn joylessness, in its ability to make the dead thing dominate the environment. This domination is temporary. It endures only in virtue of the image. If you search for this place on Google Maps and go down to street level view,[2] *you will discover that the death registered here now hangs between two viewpoints. 'A photograph is both a pseudo-presence and a token of absence'.*[3] *For the moment, at least, the tree is both there and not there, depending on where you look. As you approach, there the tree is on the right in its field on the edge of a turning. 'Advance' with a click along the virtual road to the corner of the turning and the tree is gone. Whither memory?*

SECTION 1: PUNCTURING LIGHT

Nancy M. Shawcross points out that Roland Barthes' writings seem to invest photography, the quintessentially modern medium, with a peculiar mythical power. This is typified by the moment when, 'at the end of [the essay] "The Photographic Message", Barthes links the "pure" imagery of the photograph to nonculture, while maintaining that the visual messaging of a drawing, for example, can never elide culture'.[4] It seems stretching credulity to claim that an image, just because it is reproduced mechanically, occupies a position beyond culture that one can access just by looking at it; that a place beyond conventions exists which can reach out and touch one 'directly' via the optic nerve. But Barthes seems obstinately to cleave to the notion that the photograph marks an essential historical moment which breaks with everything about the image that has come before it. This is memorably expressed, as Shawcross notes, in an interview from 1980 in which Barthes comments that the forthcoming text of *Camera Lucida*, which offers what must be his most personal and provocative ruminations on photography, is concerned with elaborating the implications of the photograph as 'a new type of image' whose very newness empowers him. In viewing such images, Barthes says, 'I place myself in the situation of a naive man, outside culture, someone untutored who would be constantly astonished at photography'.[5] The photograph, in other words, can be taken as a pretext for adopting an innocent openness with regard not just to what it presents to the viewer but to the medium itself. It is a matter of attitude. Barthes *wants* to be amazed and perplexed by what he sees and makes this an essential element of his experience of photography, allowing himself to be seduced by what has been termed 'the myth of photographic truth'.[6] The technology of the image is to this degree approached in a manner somewhat reminiscent of Walter Benjamin in that it is treated as being endowed with the potential to liberate us from history.[7]

Barthes' purposely naive approach involves his responding personally to the images he sees, working above all 'by being guided by my pleasure'.[8]

This personal engagement leads to a realization of the photograph's intransigence when it comes to the individual's interpretative demands. The photographic image resists: however long one might look at it, 'the Photograph cannot be penetrated' due to the 'evidential power' it holds over us. Thus, 'however long I extend my observation, it teaches me nothing. It is precisely in this *arrest* of interpretation that the Photograph's certainty resides'.[9] In other words, a photograph when it is treated as an object of personal concern stops meaning in its tracks. With the photograph, one's search for sense causes sense to withdraw itself. Barthes' deliberately 'naive' attitude may be problematic, as Shawcross comments, especially given his deep interest in exposing cultural mythologies. But one can treat as self-consciously provocative Barthes' foregrounding of the innocent notion that the photograph has an unmediated relation to its object. Regarded in this way, his apparent naivety is deceptive: it conceals a complex attitude intent on exposing in a productive manner a disruptively mythological element that accompanies our relation to the photographic image.

The complexity and sophistication of *Camera Lucida* is revealed in its careful plotting of the range of meanings that accrue to photographs and, above all, in the text's introduction of the disturbing notion that the photograph, while usually conceived as preserving life by capturing its subjects in the midst of living, is overwhelmingly marked by the presence of death. 'All those young photographers who are at work in the world, determined on the capture of actuality, do not know that they are agents of Death'.[10] This is because photographs do not summon the past in a way that brings it back to life for us. Rather, photographic images attest to the fact that the people preserved by way of them once existed, 'that *they were there*' and in the image still abide 'there',[11] but also are no longer.[12] In the photo a person is caught alive in the midst of their impending demise.[13] What is unsettling about the mechanically reproduced image (especially when it is untainted by the process of colour reproduction—'an artifice, a cosmetic', says Barthes dismissively, provocatively, and maybe rightly) is that it conveys the impression that the dead person's 'photographed body touches me with its own rays and not a superadded light'.[14] The photograph's touch is the touch of death, a corpse's squeeze of the hand pressed into a living one by mechanical means. An image, in this way, becomes in the very moment of its making a prefiguring of its subject's demise. The photograph's subsequently being viewed consequently initiates a torn temporality—one sees those past, i.e. the dead, now as living. To see a photograph innocently as embodying a direct impression of the person it captures, unmediated by culture or interpretation, is to unleash the guilty thought that the person's recorded image thereby intimately touches the seer who, in this way, brushes up against the dead one's living presence—that the dead touch one's optic nerve directly, as it were.

This intimate contact wherein light enters the viewer's eye from something generated directly by a body once in front of the camera's lens suggests something untamed and dangerous. Barthes locates this uncomfortable occult intimacy as being central to the photographic image.[15] For this reason, social order seeks 'to tame the Photograph' either by conferring on it the status of 'art' or by spreading its products everywhere so that the medium ceases to stand out as intimating something mad and dreadful. It would be possible to follow many possible trajectories from Shawcross's deeply thoughtful engagement with the elusiveness of *Camera Lucida*. However, two related aspects stand out. These are Shawcross's elucidation of Barthes' deployment of the notion of the palinode as a means of dividing the text of *Camera Lucida* into two parts, and Barthes' conceptions of *studium* and *punctum*.

A palinode is a retraction or, as Barthes puts it, a 'recantation'.[16] It forms, Shawcross argues, the 'methodology'[17] on the basis of which the text of *Camera Lucida* is constructed. This method allows for a structure composed of points which by turns assert and then repudiate themselves in a series of pushings and pullings that play with the categories of unity and fragmentation. Barthes' purposely 'naive' approach to the topic is, to this extent, deeply self-reflexive in its repeated self-de-structuring of the text's modalities of argumentation. It is '*naive un-naive*'. Above all, this means that the central categories the text deploys of *studium* and *punctum* and its cultivation of the notion that the image has an occult quality of presence take on shifting meanings as the text's argument moves on which makes them hard to pin down.

If one wants a broad definition, the *studium* is the '*average* affect' of an image. It is linked to such things as compositional conventions (e.g. the use of the 'golden ratio' or the 'rule of thirds'). To encounter the *studium* in an image is to recognize in it what the photographer intended us to recognize. It confirms that 'culture [. . .] is a contract between consumers and creators'.[18] The *studium* is the realm of the marketplace, of satisfied customers who get what they pay for, namely, a carefully prepared 'experience'. The realm of the *studium* is dull. *Puncta*, on the other hand, consist of those things which, because they lie beyond the maker's control, defy casual consumption. A *punctum*, as its name implies, is what punctures conventional presuppositions, like a 'sting, speck, cut, [or] little hole'.[19] An image's *punctum* springs from photography's link with what Robin Kelsey has memorably characterized as 'the art of chance'.[20] This is the unpredictable and coincidental element in some images which exceeds the parameters of technical control, takes hold of the image, and comes to dominate the viewer's perception of it. The realm of the *punctum* is surprise. *Puncta* are 'at once brief and active',[21] like little explosives. They do not conform to a pattern. Very often they are small details such as the coincidental aspects of a photograph, accidents of chance—the strange angle of a person's hat; the way a shoe is laced (or remains unlaced); a figure that is partly obscured by the image frame; a side-long, furtive glance. They are disruptive since their presence does not accord with compositional conventions or the intentional deliberations which tend unconsciously to follow them.

Both *studia* and *puncta*, Barthes holds, are present alongside one another in photographs. But already, as soon as he deploys them, the meanings of the terms begin to shift because of the assertion of their close proximity to the realm of the personal. One cannot define in advance how one might feel about a detail. One can, for instance, enjoy or be indifferent to *studia* and only *puncta* can be loved, but one cannot say of anything, 'This is universally lovable'. Barthes in this moment transforms the sense of these notions by elaborating them according to personal reactions, which are subjective and do not therefore follow predetermined rules. What is lovable in a photograph is not a matter of its picturesque or well-made qualities but of its surprises, of its hidden content which frustrates all technique and appears only in the eye of the beholder who must then communicate how this or that detail that has come upon them alone severs itself from the public realm of convention.

Puncta are unstable, antisocial. If one were to think of the photograph as an enclosure of 'things', then an image's *punctum* cracks open the enclosure, undoing the image's tendency to formal, rule-like containment. What one sees in a look, in a gesture, in moments caught 'between things' unleashes the photograph's dangerous potential to shock one out of the mythological faith in stable and enduring conventions which are the guardians of social order. Since it is only seen by the beholder, the *punctum* in an image exists in a twilight plane that lies beyond mere method, for it strictly neither is nor is not: 'it is what I add to a photograph and [yet it is] *what is nonetheless already there*'.[22] And it is in this way that the *punctum* becomes also a signifier of the temporality of the image (the image as the presentation of what-has-been) and the thought of death.[23]

It is not difficult to acknowledge the power and appeal of Barthes' contention (as already said, most likely deliberately provocative and semi-mythological) that there is a feeling of immediacy linked to the idea of light bouncing off a person, entering a lens, and being focused by it onto a passively receptive light-sensitive surface. It is not difficult to acknowledge the power of this idea because we are all prey to it. The idea guides how we read photographs: we presuppose authenticity of this order whenever we view a photographic image as if it provided testimony to events by passively replicating them as if we 'were there' (we cleave to faith in the camera as objective witness to events because we cleave to the dubious belief that we, too, can stand in an immediate sensory relation to things. This is one of the pragmatic conditions of our relation to the world of practices we inhabit: the exposure of deceit in a photograph makes us uncomfortable, not least perhaps because it places this presupposition of authenticity, of our trust in our eyes, in question[24]).

We tend to (want to) believe the photograph. To consider the extent that this is so, imagine if a hitherto undiscovered legacy of images made by a photographic process from the era of the ancient Greeks were to turn up. It would be difficult not to place more of a burden of reality (of trust) on photographic images of Athens or Sparta, or photographic portraits of Aristotle or Thucydides, than one would ask to be borne by any drawings or sculptures. (What if such photos contradicted some cherished things we might believe about these places or people?)

That such faith in authenticity is bound up with morbidity, as Barthes argues, may strike someone as being rather more alien to the above presuppositions, but Barthes has a point when it comes to the photographic medium of analogue film. The presuppositions of immediacy, the fixing of the slippage of time and mortality, are linked. The relation turns on the assumption of presence. The film which holds the image has, we feel compelled to believe, to have been 'there' physically in the historical environment of the subject or thing captured in order for it to record them (even if it is situated at some physical distance in virtue of being copied by way of a printed negative). That the plane of the photographic plate or film is flattened out to two dimensions, that it is already organized according to conventions of representation and proportion, and that a lens must also be there, too, a lens which bends light in the manner peculiar to it so as to form the image on the sensitive surface, does not undo this entirely even if it from the outset renders the supposition of an unmediated relation deeply problematic. (A lens, after all, is a choice made for directing light in a certain way—its angle of view includes or excludes things, exaggerates things or makes things seem 'natural'. A chosen lens's aperture, too, is selective, allowing foreground or background to be washed out in a blur or sharply drawn into the plane of the situation it thereby constructs. There are also issues of exposure and lighting, of framing and the accidents of processing which haunt the production of film and print.)

A hypermodernizing problem, however, arises as soon as we commit the sin of going digital. As Christopher John Müller and Ian Wiblin note, the near infinite manipulability of a digital image means that 'today the photograph is no longer the placeholder of something that necessarily exists' for it has become 'a pixelated mosaic that can be altered at will or built up from scratch'.[25] Thus, an ontological and epistemic gap opens between the viewer and the viewed which for a brief moment of Barthesian rapture might have been assumed to be bridged by the medium of film.

The problem, however, goes even deeper than the loss of the 'direct, physical relationship to a continuous pattern of light generated by objects'[26] that film might be assumed to have. The digital photographic image is not only a prey to the alteration or formation at will of pixelated patterns thrown onto a sensor by a camera lens. Its treatment of light is also subject to a kind of intervention that is independent of any manipulation that a person armed with a photo-editing programme might inflict upon it before or after the fact, for what is at stake is a new kind of interpretative automaticity related to the near 'real-time' relation the computing device has to its environment. With digital photographs there is a deeper and more profound loss of the possibility of a 'one-to-one correspondence with what they show'.[27] It is a loss that extends through the totality of the digital imaging process and cannot be side-stepped as it is a necessary condition of its possibility. This is because digital imaging works by way of processes of computation consisting of procedures which even *precede* the arrival of light conveyed by a lens. Where a piece of sensitized film consists of random arrangements of concentrations of suspended silver halide crystals awaiting the exposure which will make them resistant to being chemically dissolved by a developer so they can then be hardened in place by a fixer, the digital sensor consists of a coordinated and fixed arrangement of microscopic wells containing light-sensitive pixels each of which can generate a signal that is fed into a computer in the form of electronic information. This information is then amalgamated to make the image. In the case of CCD (charge-coupled device) sensor cameras, this means taking the analogue charge provided by the sensor and amplifying and then translating it into a digital form that can then be stored on a memory card. In the case of the now commonly used CMOS (complementary metal-oxide semiconductor) digital sensor, this process begins with the sensor itself: each pixel is linked to a solid-state circuit on the sensor chip which manipulates the data 'on-site' prior even to its being sent to the external processor situated within the camera body, which makes it information ripe for storage. The technicity of the camera-computer (its automated, mechanical calculation and organization of electronic elements) is 'there' waiting before the image is taken. *Even before it appears*, the resultant photograph is, in this regard, an active mechanical interpretation of distributions of energy. The subject, the photographer, has no active role in this.[28]

An interpretative gap therefore opens between the image and the referent that cannot exist in the analogue process (even if one might suspect film of interpolating other things) because of the role of the electronic 'interpretation' that is engineered into the digital imaging device's computational abilities. This gap cannot be bridged even by the suspension of disbelief which might accompany film once one knows of it since the digital image is always at least as much a computer interpretation as it is a 'copy' of what the lens presents to it. This kind of intervention defies all individual 'technique', for it occurs even before the camera gets to be pointed toward an object and must consequently be dealt with by technique after the fact. We do not, in this regard, ever engage with a photograph as a pure and simple replication of distributions of light since, in the words of Eric Fossum, one of the developers of the CMOS sensor, 'some of the things we associate with photography [and photographic images] are artifacts of the way we capture them'.[29] The intercession of the computer's iteration of what the sensor receives begins the digital photographic process as much as it concludes it with a 'finished' image that is always open to further manipulation. This digital interpretative gap is what allows the image to be stored as information that can be subsequently split into 'layers' which can then be manipulated individually and laminated together ('merging') via a photo-editing programme.

The fluid elements which characterized the photograph of the film era (the fluidity of coatings of light-sensitive materials and the fluids of development) give way to a new kind of fluidity of energy. What might this hypermodernizing liquefaction of the integrity of our relationship to the image imply? One way of responding to such a question might be to follow Barthes' example and approach the photograph in a manner that is purposely 'naive' and personal.

SECTION 2: ANALOGUE INNOCENCE

My first photographs were taken with a 1927 Kodak Box Brownie (probably a No. 2 Model F). It cost half an old penny and came from one of the numerous jumble sales which at one time regularly took place where I grew up. Each of these events was a chaos of fascinating and less than fascinating junk piled up on benches in a village or church hall. In such moments, up-surging consumerist frenzy collided with the remnants of post-war austerity. People who had once known what it meant to 'do without' before times of plenty and still remembered this used these sales, as buyers or sellers, to recycle once new but now outmoded things for cash and charity. There were old 405 line monochrome televisions, wooden-clad and Bakelite-shelled radios, model aircraft, heaters, fans, vacuums, convection heaters, toys, board games and jigsaws with pieces missing, clothing, diminished sets of crockery, cutlery, broken and functioning clocks, watches . . . almost anything. Later, close friends now vanished into the divergent paths of adulthood would seek out these increasingly rare sales of jumble for electronic spares, hunting for a defunct television to take home and see if it could be repaired before stripping it and looting its magnets and other viable components in order to put them to use for some other pet project. The occasional high-voltage mishap did not seem to bother any of us.

Before the time of such lootings there was that Box Brownie. It was covered with fake leather ('leatherette') and exuded that strange quality, detectable only by the still nearly very young, which comes from objects of seemingly infinite age. It was, after all, over forty years old—a year younger than my mother and thirteen years younger than my father, and it seemed an ancient, mysterious enclosure to my seven-year-old self. The camera had a double viewfinder (for horizontal and vertical), a shutter lever, a winder, and nothing much else. It was bulky, with a carry-strap for sliding the hand under. The Kodak 120 film took eight exposures that could be counted by numbers that appeared in a small red window on the back as the film was wound on. I remember being struck by the fact that it would wind in only one direction, forward to the next image, and that because of this a certain relation to time was already engineered into the object. After the first time I used the camera, on a seaside holiday, I discovered by accident that this relation could be frustrated by discharging the shutter more than once without winding on, allowing two or more separate sources and situations of light to exert their different influences on the film (a 'double exposure')—a mishmash of places and moments. Nearly half a century later, at least one holiday image survives: a photo taken from behind of one of my brothers sitting, half leaning out of an open car door, one leg on the ground, looking away. 'Don't', he said. *Click.* 'You didn't?' I had. He had lifted his hand behind his back and raised two obscene fingers in an attempt to stop me taking the shot at the very moment I took it.

The real magic of the experience was not entirely the transgression involved in the taking of people's light and the light of things. It was also the subsequent wait and envelopment in darkness necessary to complete an image's path into the world. I was shown how: a developing drum, the exposed film wound in at night away from the light; the pungent chemicals and their dilutions; the rinsing and drying of processed negatives; then a large glass plate torn out of an old car door placed on top of them with printing paper below; a brief exposure from a bulb; an image appearing on paper floating in trays; fixing; more hanging up to dry. The finished image (just over two by three inches) was heavy with the other-worldly, un-domestic, transfiguring odour of industrial processes. It was also resplendent in its offensiveness: my brother, about sixteen years old, gesturing grossly from his half-seated position in the car.

The relationship to technology of the relative innocence of childhood is one of a strangeness which is as yet also unafraid, a becoming-familiar. There is experimentation which preserves itself in the negative's accidental consequences of playful ambition. Consider a 35mm frame from my early teen years which also bears the scars of its ageing and careless storage:

With such playfulness there is no idea of the uses the photographic image can and has been put to: reconnaissance for bombing, staid poses in the sterile studio, 'historic' moments, the dead in death, and always everywhere the living before death.

Naivety, however, cannot last for long. I remember this first striking me when, around ten years after the Box Brownie, I was at an auctioneer's sale. There were cameras I could not afford, and as I perused them a man stood by me and began to lament the lost joys of photography. Having become a police photographer, he had spent thirty years capturing scenes in emptied houses, by closed roads and deserted, fenced-off woodlands where grief had visited someone and left only foetid, rigour-mortised remains. He offered me a photographic 'Don't' by way of advice. 'Whatever else you do, never be a police photographer', he said. 'It blunts creativity'. Picking up a camera, all he could see was his job: 'You can't forget, so you block it out. But if I try to do anything, like a landscape, in my mind's eye the scene always ends up hiding a body behind a tree and being composed around it. I could never do a portrait. It always looks like work. And now I don't even work'. He had discovered the Barthesian relation to the image as one of death in the most brutal terms and attested to its unforgettable truth.

Loss of innocence comes from connection. With it we move from the productive waste of play to the knowing squander that tends to see only disguise.[30] The more one connects, the more everything is implicated in everything else. There is less free space for playfulness, for the happy accident;

more for the unhappy insight that comes with awareness of the more sinister possibilities of use. The retired police photographer knew that. He had come to see photography, which initially he loved, not as the uninvolved and innocent recorder of accidents and deeds of violence, as distinct from his joy of creation, but as their intimate regulator and accomplice—as the disseminator and co-perpetrator of violent acts. This was visited upon him 'photographically' in the unwelcome synthesis of visual prosthesis, light emitting bodies, and grim paid labour. In this synthesis, the conjoining of tools and eyes, movements of the hands, placements, angles, and light, the things to be recorded had all come together to form the memorial death shroud that he saw as amounting to his professional career's destruction of his photographic innocence. How could he, after that, ever take an 'innocent' landscape, one not populated by trees of death?

SECTION 3: DIGITAL'S IN

If, as Barthes' argument suggests, the retired police photographer's experience represents a kind of norm and any photograph of a person is accompanied by the pricking threat of an awareness of their impending demise, since the 'who' of the person will in the end be drawn inexorably down into that mummifying, flattened plane of the visual remnant, then one is entitled to consider whether the digitized image might be more or less susceptible to this condition. The inherent element of manipulation which characterizes digital (from beginning to end) might, after all, be said to break this direct link beyond doubt. At the same time, it might also initiate death of another variety. A kind of demise of the subject is enacted by digital that analogue film never could achieve insofar as the digitally encoded light of a person is from the outset no longer his or her living remnant but only a deathly afterglow and trace of invisible interpreted energy. In this regard, digital might universalize a deathly element, albeit not as a celluloid fixing of time but rather by dissolving the person's presence through making it *liquid* transferable electrical 'information'. Under such a regime, 'we' would live on only as already interpreted images of resurrected traces of data, as light-ghosts whose digital remainders are translated into information and stored in the prosthetic memories of computers: 1s and 0s awaiting twisted resurrection in the manipulative luminescence of the screen.

Digitization certainly makes budding police photographers of us all in one regard. We are increasingly assiduous makers and cataloguers of the image, of our digital self-image, in a manner and on a scale never before the case. People today seem to be becoming the recorders and gatherers of their own obituary artefacts in an unparalleled way. Every innocently uploaded selfie added to the public space of the Web waits to become a mug-shot or epitaph for curious public consumption should tragedy swallow someone up, as sooner or later it does. Each video recording is a potential archive (moving and still) awaiting upload. If a radical, exhausted disillusionment were to follow from the over-exposure to the image which is currently the norm, if we were to stop recording ourselves incessantly, this situation might change. But the continuing rise in the mass production and consumption of digital photographs today suggests disillusionment is unlikely.[31] We are and will surely remain image-addicted. The consequence of over-exposure is, however, paradoxical. To the extent that there is a universally casual acceptance of the mechanically reproduced and manipulated image that hypermodernizing digitization compounds, the deathly thought that might provoke anxiety in the photograph becomes tamed and denuded of its power to shock. Its deathly quality recedes with its normalization.

In fact, such a situation is precisely what Barthes foresaw and dreaded in relation to analogue photography: 'This is what is happening in our society, where the Photograph crushes all other images by its tyranny: no more prints, no more figurative painting, unless henceforth by fascinated (and fascinating) submission to the photographic model'.[32] The photographic model, when it becomes a tyranny, naturalizes the death it opens us to and thereby sanctions it, blocking from view the 'mad' element that Barthes identifies as springing from mechanized vision. With digitization this tyrannical model inheres even as it changes in at least one significant way. The model itself is a blend of past conventions and 'modernizing' uses driven by the commercially dominant command that we must 'communicate' and show ourselves (since this 'must' is a source of revenue, a contemporary seam of gold). There is still the usually rectangular frame and its ratios, the mode of composition which seizes the eye as much when it is defied as obeyed, an enclosure of space which communicates by simultaneously containing and excluding what may or may not be 'there'. There is also the pointing of the lens which absents itself from thought in directing the attention to that which it points at—an absenting which is a residue of the old Western inclination to see in terms of transcendence. By making the model of the photograph and these aspects natural, normal, everything photographed becomes threatened with being 'naturalized' and normalized, too, as a mere object of curiosity.

Death is no exception to this rule. When we look at death in digitized form we tend to see something that 'happens' more often and in more places, yet all mediated by the one non-place of the digital screen which saturates and conjoins elements. Death happens as digital-image-at-a-distance, as 'representation' becomes 'information' to be pondered alongside the plenitude of other images scattered about in newsfeeds and websites. Bodies and weeping after bomb blasts, machine parts and body parts scattered across landscapes, aircraft piling into buildings, 'surgical strikes' nestle up and meld with pictures of kittens and advertisements. It matters less and less if they are 'real' or not. These 'representations' all mutate through digitization into the levelled-out easefulness of *entertainment*, i.e. something wretched only insofar as it is watched casually, in the appropriate attitude of 'horror' and outrage or vague amusement, and then largely forgotten, especially in terms of any relation that might pertain to the everyday world one is inclined to accept without question. In all such cases, a spectacle is offered which the digital image conveys and in the very moment of conveyance simultaneously withdraws itself from the gaze. The role of the image and the electronic dissemination inherent in the interpretation of spectacle is concealed. This does not mean that digital transformations are less authentic and less capable of telling us about the subjects they depict than analogue was. But 'saying something about' is not recording. It is interpreting. The interpretation tends to be continuous too: once taken and saved, the image lives on mostly on the animated screen rather than within the relatively solid layers of printed materials. The digital image's materiality is now the materiality of electronically organized arrangements of light which must continually act in order to 'reproduce' it.

With all this, the production and reproduction of the digital photograph has become *as much* a matter of manipulative technique as that of any image made by drawing or painting (which have also moved closer to the digital photograph by becoming digitized too). This suggests that all the depicted elements in front of the digital sensor are temporally subordinated to interpretative processes in a manner that film could not be (such as the use of in-camera 'effects' like the manufacture of 'bokeh'[33] by some smart-phones). Digital processing now precedes the seeing it is meant to facilitate. This initiates that loss of visual innocence that one can call 'digital sin'. It is also an 'unoriginal sin' since it manifestly dissolves the origin as unique referent of perception in a manner that no film camera could. Digitization renders every image not merely an analogue copy of something but an always already manipulated (in-camera) approximation of the lens's play of light, an electronic *interpretation* of the mechanical interpretation inherent in the less-palpable material elements that make up the digital camera and span the photographed subject. Like the retired police photographer, once one notices this one cannot help but ponder its implications for every image even in the making. Every tree is a potential tree of death, however burgeoning with life, for it is already colonized by the digital artefact.

A general unease about sinful digital manipulation finds expression in a recent resurgence of interest in the non-digital camera and talk of a 'return' to analogue.[34] But this return, even if it were to turn out to be a widespread phenomenon, could never close the shutter on digital saturation—the liquefaction of light which the in-camera computer performs independently of the photographer—since all images, if they are to be widely disseminated, must now be displayed on the Internet and hence digitized. To this one could add that what is saturated cannot in any case be easily overcome since it is too fluid and malleable to be threatened all at once everywhere. Once it has become digitized information, no image is ever entirely discrete or can be viewed as such. And since there is no image that cannot be digitized, not one is immune from the suspicion that it might have been already. The film image, too, is 'digitized' by the arrival of digital simply in virtue of the suspicion that it may have been scanned and so digitized. Digitization is, in this regard, as much a cultural as a 'purely technical' event in its focusing of the vision.

This change is not all modernity in accelerated form. The universal digitization of the photograph initiates a strange, archaic transformation of our relation to what we 'see'. This transformation has two significant aspects. First, the very idea of the photograph as a maker of analogue forms whose sources reside in what emanates directly from the things photographed is rendered out of date and archaic in the sense that its original mythical power as prime seizer of light is forever lost to us. Modernity's moment of analogue innocence was, it follows, a brief interlude—like all innocence, a mere prelude to sin. After this loss, one can no longer be unknowing about the relation between the imaging device and what is seen since there is manifest construction and manipulation involved at every level that transcends the 'mere' subjectivity of the image taker and whatever he or she might decide to point the device at.

Second and more strikingly, as has just been hinted, the unlimited possibilities of digital manipulation mean that the photograph now has begun to share something in common with the archaic forms of image making it was once supposed to have outmoded and surpassed; namely, those of drawing and painting. As William J. Mitchell comments, with digital we encounter something that asserts itself by 'differing fundamentally from both photographs and paintings'.[35] However, by way of this difference less of a gap comes to separate the digital image from the cave painting than separates it from the analogue film image whose progeny it is. If a digital image is manipulated at the moment of its taking and can be subsequently and endlessly reinterpreted and 're-painted' in a manner invisible to any subsequent viewer (appropriately, photo-editors are equipped with 'brushes'), then it is no longer in any possible sense a passive re-presenting of the light that fell upon it. The digital image has lost photography its assumed (and from the outset questionable) authoritative status as an 'objective' receiver and recorder of a light of objects that is simply 'there'. The hypermodernizing digital photograph can no longer even so much as *hope* to be, as Barthes put it, a pure mechanical denotation unsullied by interpretation. The in-camera computer which extrapolates and estimates what is missing from the light sensor's offerings sees to that from the outset; the manipulative possibilities that ensue for the camera-user to play with and exploit after exposure only serve to make doubly sure. The hypermodernizing digital image, for all its technological facility, has consigned the photograph to the status of light painting. Consequently, digital photographs cannot be 'truthful' any more than the most flattering of expertly painted portraits. Digital sin, in other words, is a living, modern-archaic, hybrid condition. Its hybridity is shown by its ability to throw our trust of what we see backward, propelling us to enter a parallel state of unbelief that rises up alongside our technological faith in mechanical representation.

The digital image has the intimate relation with the tree of death with which this discussion started of being the harbinger of the demise of our faith in the medium which generates the world of the seen that we now inhabit so concretely and rely upon so much. 'News' is now 'information' too—easily manipulated. 'Fake news' is a term whose power plays precisely on the loss of faith that is exemplified by digitization. With digitization, our faith in the image's fidelity fades.

Barthes talks of once being fascinated when coming across 'a photograph of Napoleon's youngest brother, Jerome, taken in 1852 [. . .] I realized [. . .] with an amazement I have not been able to lessen since: "I am looking at the eyes that looked on the emperor"'.[36] With digital approximation those eyes withdraw their gaze from us into a new distance of death. One might say that it is no longer the subjects depicted in photographs that are permeated by death. It is the medium itself: photography as Barthes conceived of it has, for the most part, died. It lives on as a digital remnant. This does not mean that the likenesses we see spread about via the Internet are not likenesses. They are just that: likenesses. Images of this order cannot be guaranteed to be more 'representative' of 'real' life than the notorious portrait of Anne of Cleves painted by Holbein for Henry VIII was. Henry presumably looked to the painting as a source of more than mere 'information'. The life and death digital images communicate has become abstract, universal 'information'. One cannot stare information in the face.

In hypermodernization we dwell, for the time being at least, in a hinterland. On Google Maps the tree of death with which I began remains, for the moment, both there and not there. No one can look directly at it.

Who could not now be tempted to see such catastrophic manipulation in every image . . . ? There's a nuclear holocaust and desperate hopeless prayer over the dead lurking somewhere in every shot:

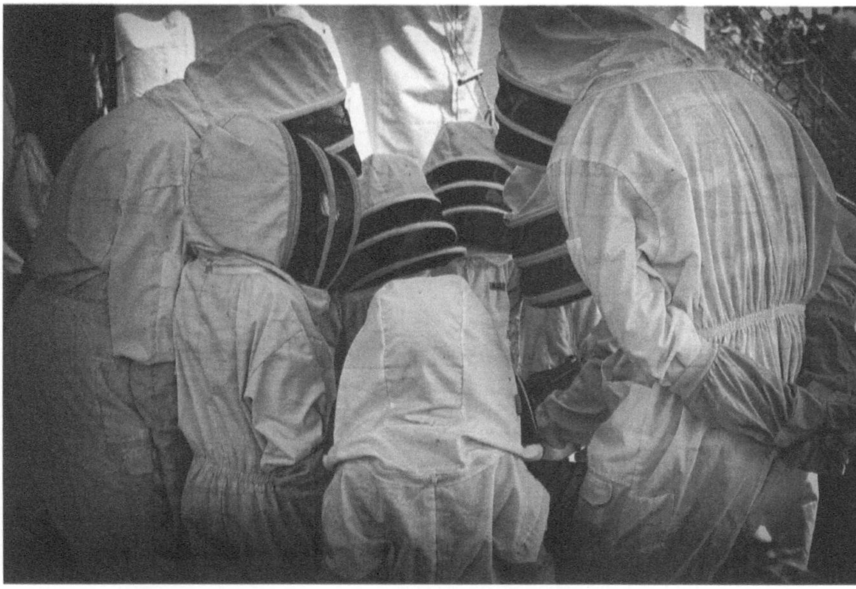

. . . even, as here, in the image of a group of beekeepers teaching children at a summer fair . . . Perhaps that means that the real challenge of digital is not a merely technological one. It propels us back toward the question of our honesty, toward 'truthfulness' rather than 'truth'. Perhaps the main question digital asks of us is ethical, therefore.

NOTES

1. 'As buds give rise by growth to fresh buds, and these, if vigorous, branch out and overtop on all sides many a feebler branch, so by generation I believe it has been with the great Tree of Life, which fills with its dead and broken branches the crust of the earth, and covers the surface with its ever branching and beautiful ramifications' (Charles Darwin, *The Origins of Species*, ed. J. W. Burrow (Harmondsworth: Penguin, 1976), 172).

2. www.google.co.uk/maps/@51.4258928,-3.5187622,3a,75y,156.84h,75.9t/data=!3m6!1e1!3m4!1szZ0Ria1LDcPZvDo3wi7HOQ!2e0!7i13312!8i6656.

3. Susan Sontag, *On Photography* (Harmondsworth: Penguin, 1977), 16.

4. Nancy M. Shawcross, *Roland Barthes on Photography: The Critical Tradition in Perspective* (Gainsville: University Press of Florida, 1997), 27. The essay in question can be found in Roland Barthes, *Image, Music, Text: Essays*, trans. Stephen Heath (London: Fontana, 1977), 15–31.

5. Roland Barthes, *The Grain of the Voice*, trans. Linda Coverdale (New York: Hill & Wang, 1985), 357.

6. Marita Sturken and Lisa Cartwright, *Practices of Looking: An Introduction to Visual Culture* (Oxford: Oxford University Press, 2001), 16.

7. See Walter Benjamin, 'The Work of Art in the Age of Mechanical Reproduction', in *Illuminations*, trans. Jonathan Zorn (London: Pimlico, 1999), 211–44. It is Benjamin who first seems to notice the historically disruptive power of the phenomenon of 'mechanical reproduction' in relation to vision. The mechanical reproductive power of the photograph, he holds, has two aspects. On the one hand, the reproduction of an artwork or a scene is capable of showing things that escape our everyday vision. Second, the ability to reproduce images and sounds changes the relation between objects and space: 'The cathedral leaves its locale to be received in the studio of a lover of art' (214). The mechanical recording of images also reveals what is concealed in our everyday activities and the spaces we casually assume in virtue of them. For example, simply reaching out for something involves an 'unconscious optics' which only the camera exposes (230). Mechanical reproduction, in showing this, demythologizes—in relation to art, for instance, it denudes the work of its 'aura' of mystery. Art in this way emerges as a site of political struggle.

8. Barthes, *The Grain of the Voice*, 358.

9. Roland Barthes, *Camera Lucida*, trans. Richard Howard (London: Vintage, 2000), 106.

10. Barthes, *Camera Lucida*, 92.

11. Barthes, *Camera Lucida*, 82.

12. John Berger makes a similar claim: 'Unlike any other visual image, a photograph is not a rendering, an imitation or an interpretation of its subject, but actually a trace of it' (*Understanding a Photograph* (London: Penguin, 2013), 51). 'Photographs are relics of the past, traces of what has happened' (*About Looking* (London: Bloomsbury, 2005), 61). For Berger, the photograph in this way presents a challenge to our understanding of the present and hence ourselves for it possesses redemptive potential: 'if the past becomes an integral part of the process of people making their own history, then all photographs would acquire a living context, they would continue to exist in time, instead of being arrested moments'. The death inherent in the image that Barthes identifies might, it follows, be a symptom of the repressive forces which structure contemporary social order. These forces, for Berger, are those of modern capital, which generates a distinction between an image's private and public use. Private use is personal, significant, imbued with the sense of being animated by the living narrative of those who in seeing images see themselves implicitly in them. Public use tears the image from its living context, rendering it 'a dead object'. For Berger, a collapse of the public-private distinction would threaten to overcome this, re-animating the image by making its significance concrete. In the era of hyperconnectivity, one might suggest, this overcoming of the difference between public and private turns out to have more unsettling implications insofar as it enacts the dissolution of the distinction through the colonization of private space by subjecting it to the power of automaticity.

13. An understanding of this possibility of the photograph is clearly evidenced by Hippolyte Bayard's 'Self Portrait as a Drowned Man' (1840). See Juliet Hacking, ed., *Photography: The Whole Story* (London: Thames & Hudson, 2014), 20.

14. Barthes, *Camera Lucida*, 81.

15. According to Kaja Silverman, the assertion of this intimacy is one half of the 'pitched battles between those who champion the evidentiary value of the photograph and those who emphasize its constructedness' (*The Miracle of Analogy* (Stanford, CA: Stanford University Press, 2015), 17). Barthes' attitude, Silverman adds, reflects an attitude of 'political despair [. . .] our sense that the future is all used up'.

16. Barthes, *Camera Lucida*, 60.

17. Shawcross, *Roland Barthes on Photography*, 75.

18. Barthes, *Camera Lucida*, 28.

19. Barthes, *Camera Lucida*, 26–27.

20. Robin Kelsey, *Photography and the Art of Chance* (Cambridge, MA: Belknap Press, 2015). Kelsey notes that William Henry Fox Talbot (inventor of the first viable negative photographic process) grasped well the appeal to the Victorian public imagination of the role of chance in the making of a photographic image. On the one hand, Victorian attitudes stressed the value of labour and experimentation as part of the demanding process of self-improvement. On the other, when Talbot accounts for his image *The Open Door* (1844) in terms of serendipity, he deploys a narrative in which 'the aesthetic potential of the scene supplants rather than inspires labour' (19). Such supplanting is presented as intrinsic to the appeal of the photographic process for it removes the need for the effort and investment of time which drawing and painting require. In appealing to these two aspects, Kelsey points out, Talbot revealed how he 'understood and exploited the inconsistency between espoused morality and the emergent consumer economy' (20).

21. Barthes, *Camera Lucida*, 49.

22. Barthes, *Camera Lucida*, 55.

23. Barthes, *Camera Lucida*, 96.

24. The case springs to mind of the winner of a Nikon competition in 2016 (the prize was a 'Nikon trolley bag') who pasted an image of an aircraft passing overhead into another image of a ladder in a metal cage running up the side of a building. The manipulation was sufficiently clumsy as to be obvious to many who looked at the image, and much mockery and complaint ensued (Elsa Vulliamy, 'Nikon Photography Competition: Winning Entry Turns Out to Be Badly Doctored Image', *Independent*, 1 February 2016, www.independent.co.uk/arts-entertainment/photography/nikon-photography-competition-winning-entry-turns-out-to-be-badly-doctored-image-a6845196.html).

25. Christopher John Müller and Ian Wiblin, 'The Bank of England in Ruins: Photography, Money and the Law of Equivalence', in *Credo Credit Crisis: Speculations on Faith and Money*, ed. Laurent Milesi, Christopher John Müller, and Aidan Tynan (London: Rowman & Littlefield, 2017), 114. Müller and Wiblin are picking up on a point made by William Mitchell, for which see William J. Mitchell, *The Reconfigured Eye: Visual Truth in the Post-photographic Era* (Cambridge, MA: MIT Press, 1992).

26. Gillian Rose, *Visual Methodologies* (London: Sage Publications, 2016), 6.

27. Rose, *Visual Methodologies*, 6.

28. This is the case most obviously with the commonly used JPEG images produced by the in-camera computer. It is, however, also the case with so-called RAW

images which have not been subject to this second-order stage of processing. The RAW image, in order to be stored, is already an interpretation of light rendered as transferable data which, in order to be viewed, must be interpreted by a device operating according to rules stipulated in advance.

29. '"Perfect" sensors might be possible but might not come to cameras' (Richard Butler, *Digital Photography Review*, 28 February 2018, www.dpreview.com/interviews/9385721576/perfect-sensors-may-be-possible-but-might-not-come-to-cameras).

30. William Blake, 'The Nurse's Song', in *The Portable Blake*, ed. Alfred Kazin (Harmondsworth: Penguin, 1979), 106.

31. In 2000, the year of the height of film-camera use, around 80 billion photographs were taken worldwide. By 2015, this number had been dwarfed by digital imaging: an estimated 1.3 trillion photographs were taken, around 75 percent of which were taken with phones (Stephen Hayman, 'Photos, Photos Everywhere', *New York Times*, 29 July 2015, www.nytimes.com/2015/07/23/arts/international/photos-photos-everywhere.html).

32. Barthes, *Camera Lucida*, 118.

33. 'Bokeh' is a common word for the out-of-focus effect that some lenses will produce. One common use is in portraiture to isolate the subject from his or her background.

34. This is a recent Kickstarter-funded reflex SLR film-camera which proclaims itself as initiating a return to 'analogue photography in the 21st century' (Robin Stummer, 'Back to the Darkroom: Young Fans Reject Digital to Revive Classic Film Camera', *Guardian*, 28 January 2018, www.theguardian.com/uk-news/2018/jan/28/does-reflex-slr-camera-herald-35mm-film-renaissance). Other 'retro' film cameras exist or are rumoured to be in production.

35. Mitchell, *The Reconfigured Eye*, 55.

36. Barthes, *Camera Lucida*, 3.

Chapter Four

Speed of Tragedy . . .

Speed Camera

The Eye of Horus is put out;
above the pyramid on dollar bills
the Eye of Providence clouds to cataract.

Our own gape
ousts the old gods' stare.

We are each other's watchers,
our own worst www.walsinghams.

Pore over this (self-)portrait –

the car, in blurred italics, bent
on escape, accelerating back to sepia,

but all the more detained,
its penalty fixed
in its own speedometer's

tearless gaze.

The greater the speed the less clear things become, which is why the eye stands out, problematic, unreal. Acceleration forces confusion. This is because in an ever faster world there is always more that can happen in a given time. A span of accelerated time is not, in this regard, like the span of a poem, which can be returned to and pondered in a manner which allows language to reflect upon its own painful red births in blood and the primitive belly. It is, rather, a span which rushes past oblivious in a primitive delusion of sophistication that still cleaves to its faith in the hubristic magic of the will's imaginings. Its passing is a passing of steel through a globe turned to jelly. It is self-blinding. The 'new' grand universal has arrived and, like a monarch, it sweeps by along its royal road, doomed as of old to undo itself by way of its own troubled striving for light.

SECTION 1: SPEEDING PAST OEDIPUS?

Think of Oedipus . . . So many things seem to so many people to be 'Oedipal'. As David Kovacs notes,[1] *Oedipus Tyrannus* has been taken as a justification of punishing a flawed character or an illegal murderous act,[2] as a tragedy of fate in which humans are playthings of the gods, or as pure art unconcerned with moral matters. To these one can add further possibilities: it is a study of the consequences of an uncompromising search for truth[3] or of the guilt of a suppressed incestuous marriage,[4] an expression of what for some might seem the harsh ethos of the Greek conception of justice,[5] or maybe a definitive self-conscious overturning of fifth-century Greek pretensions to regard human reason as the measure of all things.[6] *Oedipus* is complex, 1,530 short lines of text. The tale seems to be like an elastic skin that can be drawn over millennia to fit all. It seems to be typically classical fifth-century BC in its concerns, but the play's relevance is also quintessentially modern twentieth. Why else would so much be written about it in the modern era? To make *Oedipus* harmonize with the modern in this way, however, is to pose a problem which necessarily extends beyond the very modernity that is invoked in order to frame it as such. Sigmund Freud is one figure who resides at the heart of this problem.

Quite a few people know the story of Oedipus.[7] Absolutely everybody, however, seems to know about the Oedipus complex. In contemporary culture, one suspects, it is the complex that for most frames the mythic figure. 'Isn't he the weirdo who killed his dad and slept with his mum?'

After watching performances of Sophocles' play in Paris and Vienna, Freud made the ancient figure of Oedipus utterly modern by casting him as a psychoanalytic (and hence 'scientific') symbolic representation of the hidden forces that underlie the 'normal' everyday world of familial relations which purport to ground social order.[8] The rhetorical appeal to the authority of modern science is central to the Freudian approach. Science is method. Method matters. Method facilitates the authoritative illumination of what is hidden in the dark recesses of the mind. The methodologist is a kind of explorer armed with theory who goes in search of unknown specimens. It is a conceptual battle with the natural realm. Freud goes out to look at nature by turning inward, eyeing a structure he deems the mind's 'unconscious'. Amongst Freud's proudest finds, as Lytton Strachey puts it, is his '*discovery* of the Oedipus complex'.[9] By being conceived as a 'discovery', the Oedipal is turned into a kind of thing passively waiting to be unearthed by the proper means and baptized with a proper name. It is like a hardened fossil shell trapped in sedimentary rock awaiting a kindly geologist's hammer blows. What the ancient unfolding structure of a fossil shell reveals about living shellfish finds its parallel in what the psychoanalytically unfolded, ancient Oedipal psyche tells us about us moderns.

Psychoanalytic cracks open mental rock to reveal telling Oedipal traces lingering. We are supposedly dealing here with a pure fragment of nature that transcends cultural and historical specificity. The 'profound and universal power' of the Oedipus legend, Freud contends, reveals it to have 'universal validity'.[10] Myth, one can say, from this point of view transcends the mythical when it is seized upon by modern theoretical mastery. The figure of Oedipus is cast in a modern discourse that renders him timelessly primitive and current at one and the same moment: old and new, legend and scientific theory coalesce. The ancient Greek comes bearing a gift to be snatched with eager hands. Psychoanalytic theory is the direct beneficiary of a borrowed aura which glows off what is supposedly revealed by method to be both ancient and current in equal measure, culturally present yet ineffably and distantly natural. Theory in this way implicitly stakes a claim to a mimetic power borrowed from nature's primitive forces. It is this power which endows theoretical representations with the authenticity necessary to make them count as 'knowledge'.

It is no coincidence, in this connection, that Freud is keen to show that Sophocles' play parallels psychoanalytic method. *Oedipus Tyrannus*, he tells his reader, enacts its plot by way of processes which 'can be likened to the work of a psycho-analysis'.[11] *Oedipus* is textbook legend. It is exemplary proto-theory. In this way, the archaically primitive and the modern are envisaged as reaching out to embrace one another through the shared universality indicated by a common methodological trait. Freud's argument in support of this last claim strikes one as odd—or it ought to. It is that if *Oedipus* can move a modern audience as much as it surely must have moved audiences of ancient times, this is because there is more to the play than meets the eye. The eye must give way to a deeper meaning beneath the play's surface events. What Freud takes to be the up-front narration in *Oedipus Tyrannus* of a clash between 'destiny and human will' for him conceals something else which exists within us no less than it did in the psyches of the citizens of the ancient Greek Athenian *polis*. This is why the play exerts its extraordinarily powerful appeal across the ages.

This 'something else' is in fact two things: first, it is sexual desire; second, it is the timeless formation of that desire within the individual infant within the triadic structure of the family (the basic unit of the *polis*, if one is to believe Aristotle). This combination forms a fundamental structure. It comes down to the fact that 'it is the fate of us all, perhaps, to direct our first sexual impulse towards our mother and our first hatred and our first murderous wish against our father'.[12] This condition Freud calls 'destiny'. Sex is fate. As fate it is common to all humanity: Oedipus's 'destiny only moves us because it might have been ours—because the oracle laid the same curse on us before our birth as upon him'. With this, psychoanalysis stakes a claim to reveal a humanity that is timelessly natural in virtue of the disturbing unnatural inheritance it carries of a primordial curse. In this way, too, the spirit of scientific method proclaims mastery over myth by isolating myth's supposedly hidden rational content and asserting this content to lie at the core of mythical insight. The power of myth in this way gets incorporated within reason itself:

fate is universal and proclamation of this universality now becomes reason's right. Psychoanalytic reason in this way claims nothing less than the right to the knowledge of human destiny.

The justification for Freud's conclusions about *Oedipus* is offered in simple enough terms. Psychoanalysis's authority resides in what it sees and interprets through method. It beholds destiny in (and as) vision, which the analyst sees endlessly rehearsed in the dramas that take place on the visual theatre stage of our dreams. If we are driven to accept the idea that the first stirrings of sexual desire and murderous hatred are directed against our mother and father, that this is our 'fate', then it is because 'our dreams convince us that this is so'.[13] Dreams properly interpreted are authoritative. Correctly comprehended, they proffer visions which have hidden within them the sovereign authority usually reserved for the commands of the powerful (e.g. the monarch or sovereign body of a city): interpreting dreams involves establishing laws. Freud reinforces this dual visual-regal character when he claims, '*The interpretation of dreams is the royal road to knowledge of the unconscious activities of the life of the psyche*'.[14] The dream becomes a roadway whose converging parallel lines terminate in the unconscious Oedipal city. For Freud, the realm of the dream charts in visual-symbolic form the necessary repression of 'primeval wishes' in favour of the easier life of forgetting which poor old Oedipus, for obvious reasons, turns out not to be able choose. We who are not Oedipus are fortunate in that we have succeeded 'in detaching our sexual impulses from our mothers and in forgetting our jealousy of our fathers'.[15] We who are 'normal' have successfully negotiated the eternal pitfalls signposted on the dream road—although it does not seem to occur to Freud that people up to now, being successful forgetters, have done so perfectly well without his assistance.

One need not swallow whole this neat little piece of timelessness, or the visualization of ahistorical psychic space which it deploys by way of its conceptualization of the dream. For one thing, as Deleuze and Guattari argue, the supposedly fateful repression involved here does not concern anything more 'primeval' than the historically specific social structure of the bourgeois family prevalent in Freud's times.

> Insofar as psychoanalysis cloaks insanity in the mantle of a 'parental complex,' and regards the patterns of self-punishment resulting from Oedipus as a confession of guilt, its theories are not at all radical or innovative. On the contrary: *it is completing the task begun by nineteenth-century psychology,* namely, to develop a moralized, familial discourse of mental pathology, linking madness to the 'half-real, half-imaginary dialectic of the Family,' deciphering within it 'the unending attempt to murder the father' [. . .] Hence, instead of participating in an undertaking that will bring about genuine liberation, psychoanalysis is taking

> part in the work of bourgeois repression at its most far-reaching level, that is to say, keeping European humanity harnessed to the yoke of daddy-mommy and *making no effort to do away with this problem once and for all.*[16]

Psychoanalysis does not daringly expose primitive drives underlying normalcy and good social order. Rather, it cobbles together a tripartite structural relation consisting of 'daddy-mommy-me'[17] with which it is possible to eternalize the repressive, nineteenth-century social institution of the family, allotting it a place in the still expanding pseudo-scientific gallery of supposedly natural, inescapable 'facts' of human existence. To this extent, Freudian thought is an active participant in justifying the hierarchical industrialized society from which it springs. For Deleuze and Guattari, appreciating this implies straight away that the question of Oedipus is better regarded as a question of production. Understood in these terms, psychoanalytic thought stands revealed in its determination to make of humankind an abstracted industrial artefact that harmonizes 'ideologically' with the modern industrialized social milieu.

One might say that psychoanalysis never gets near Oedipus for it 'Oedipalizes' and infantilizes him by confining his fundamental emotional grasp of human relations to the sphere of his parents. Freud makes Oedipus a big kid trapped in an adult body who managed to top his dad and get hold of his mum like the rest of us little ones would have done had we not embraced the detachment and forgetting of the grown-up repressive social world and become replicators of the same process all over again by making our own pre-cursed offspring in turn.

Oedipus Oedipalized is an icon of modernity through and through. The rationale of this iconic status lies in the authority gleaned from its claimed connection 'to the father of the primal horde, to God and the Palaeolithic age'[18] for this perceived link acts to bind previous generations to the present in an indissoluble connection justifying the order of the present.

One might ask what, if not the destiny bestowed by nature, trundles along down the dream lanes of Freud's theoretical imaginings? Dreams, he tells us, have two important aspects. There is their 'manifest content' and their 'latent content'.[19] Manifest 'dream-content' is simply what the dream seems to be, its surface appearance. Latent content concerns its deeper meanings or 'dream-thoughts'. A dream's meaning is arrived at by analysing its dream-thoughts, which have to be decoded from the dream-content. For this task, the analyst has to be a kind of translator.

> The dream-thoughts and the dream-content are presented to us like two versions of the same subject-matter in two different languages, or better put, the dream-content appears to us like a transcript of the dream-thoughts into another mode of expression, whose characters and syntactic laws it is our business to discover by

> comparing the original and the translation. The dream-thoughts are immediately comprehensible as soon as we have learned them. The dream-content, on the other hand, is expressed as it were in a pictographic script, the characters of which have to be transposed individually into the characters of the dream-thoughts.[20]

It is hard not to detect here what seems like semi-borrowing from Nietzsche.[21] Yet it is distinctly un-Nietzschean notions of linguistic equivalence and immediate cognitive transparency rather than Nietzschean themes of power and ambiguity that set things in motion. The argument stumbles after the first sentence with the admission that what is presented in dreaming is not purely language but a sequence of signs that are in essence hieroglyphs which occupy a borderline between language and something which is given as mere image. It becomes at this moment ambiguous as to whether we are dealing here with a language or something analogous to writing or a hybrid of linguistic and non-linguistic pictorial elements. Freud wants to come down in favour of pure thought: he deems both modes of expression to have the same referent but holds the referent itself is only amenable to being identified when it is rendered in the abstract non-pictorial terminology of 'dream-thoughts'. The presupposition, therefore, is that the visual elements which make up the dreamscape are themselves already transcriptions relating to referents that can be made to yield meaning by being translated into analytical language. The translation of the dream's visual elements into analytical terms allows them to be read symbolically in order to uncover through interpretation the genuine object of anxiety they conceal. The regal activity of dream-interpretation involves making a sequence of substitutions wherein the picture elements in a dream are replaced in each case by 'a syllable or word' that represents them.[22] One might say that for Freud, the image is a passive object of thought opposed to psychoanalytic ('scientific') thinking as such, which is linguistic, referential, abstract, and active.

Visual elements, however, come back to Freud's discourse on their own disruptive terms. In a brief passage Freud tacitly admits to his reader that what goes around comes around. Dreams, properly interpreted (i.e. in the manner he extols), mark a movement of signs which takes us from crude visual symbols to an abstract conceptual thought capable of telling us about the everlasting structure of the unconscious. Yet his own approach compels him to conclude that the very activity of analysis is prone to bring about a symbolic return of analysis itself translated back into symbolic pictorial dream-form in the patient: 'It is not surprising that a person undergoing psycho-analytic treatment should often dream of it and be led to give expression in his dreams to the many thoughts and expectations to which the treatment gives rise'.[23] It is as if rational methodical analysis cannot help but generate irrational symbolic dream responses in patients in moments beyond its control.

The reason psychoanalytic treatment might inspire dreams is obvious and not very interesting. As part of the patient's experience, analysis is also prey to the psychic processes which make up that person's daily life—it will get integrated back into them and regurgitated out again in a dreary psychic cycle of dream expression. What is interesting, however, is the specific dream that, when it is communicated to him in therapeutic sessions, Freud repeatedly takes to epitomize the psychoanalytic process: 'The imagery most frequently chosen to represent it [i.e. the experience of therapy] is that of a journey, usually by motor-car, as being a modern and complicated vehicle. The speed of the car will then be used by the patient as an opportunity for giving vent to ironical comments'. Freud, writing here in the immediate aftermath of World War I, locks on to an exemplary image of modernity: a speeding motorized vehicle on a hectic journey. He automatically seizes hold of this image and locates himself and his method as the motive force lurking behind its occurrences in his patients' dreams.[24]

This identification is not a matter of this or that patient, notice, but of all patients who dream of speeding cars. If one were to stand to one side of psychoanalysis and refuse its authority, the association seems at best a contingent one. Indeed, the patients themselves clearly do not take therapy to be a rapid process characterized by technological speed but a slow and laborious one. Freud 'unconsciously' (a hard word to use since it presupposes one knows what both consciousness and everything that is related to it, but is not it, is) offers us what one might reasonably call a definitive self-visualization that invites itself to be read symptomatically. It is not a matter of whether or not fast cars 'really' symbolize psychoanalysis. Rather, this Freudian self-visualization constitutes an admission that psychoanalysis aspires to the contemporariness and power of the speeding technology that burgeons around it. What does it mean to make of such an occurrence a general rule? Simply that whenever he hears of a dream of accelerating motor-travel, Freud cannot help inserting the patient into a 'vehicle' of ambitiously speeding psychoanalytic method, casting himself thereby as an essentially modern controlling figure within the drama of the hectic dream journey. Modern science (which Freud's thought aspires to be) is denoted by technological acceleration: modernity's dream-image is a representation of technologically realized speed. Psychoanalytic method wants to emulate the car's unimpeded flow through the environment, its unstoppable force.

Notwithstanding noticing his patients' 'ironic comments' about the slowness of actual analysis and his own later description of it as such,[25] Freud here effectively admits to being possessed by the ongoing and overwhelming fusion of desire and machine that characterizes his culture and times. Faced with dreams of machines, he cannot resist the urge to see himself not merely in them but *as them*. He wants to make psychoanalytic thought as fashionable and powerful as a racing vehicle. Freud and his rational method through this stand indicted of being seduced by an image of the contemporary which exerts its force by non-rational means.

With this in mind, one can return to Freud's Oedipalized Oedipus. Freud's world is crawling with 'disguised Oedipus dreams'.[26] Just as Freud is tempted to see in every motor-car dream a symbolic displacement in praise of his own psychoanalytic technical prowess, so almost every other dream is prey to being taken to conceal an unsophisticated Oedipus image world: 'once more an Oedipus dream'.[27] That a patient will often question Freud's keenness to advance the Oedipal interpretation counts for little. The patient is not the vehicle, never mind the driver: he or she is a passenger. Patients are not Freudian rational method. They are all irrational Oedipuses and so lack the authority which comes with method's technical prowess. With regard to this matter, for Freud, patients are best regarded as wilful self-deceivers. They protest too much. They take the image and blind themselves to the dream-thoughts indicating the desires howling behind it. In denying the Oedipal they all follow Oedipus and put out their own eyes. Freud reads the close of the play as confirmation of this—yet another universal tendency. The chorus's final 'Fix on Oedipus your eyes . . . ' is a 'warning' which tells us that 'like Oedipus, we live in ignorance of these wishes, repugnant to morality, which have been forced upon us by Nature, and after their revelation we may all of us well seek to close our eyes to the scenes of our childhood'.[28] Vile nature endows vile desires, so we avert our eyes to the 'primaeval dream-material' offered by the Oedipus legend 'which had as its content the distressing disturbance of a child's relation to his parents owing to the first stirrings of sexuality'.[29] It is the primitive visual landscape of the dream pictograph which here is held to connect ancient and modern humanity, for in modern life 'just as then, many men dream of having sexual relations with their mothers, and speak of the fact with indignation and astonishment'. Every 'Oedipal' dream is the *same* dream, an endless repetition of unity. Freud, in claiming to have opened his eyes, however, closes them to the technological imperative driving him to the psychoanalytic construal of the relation between theory, dream, image, and myth. The construction of purportedly autochthonous, unchanging sentiments is simply another modern means to the affirmation of the technical and its incestuous dalliance with its child, method.

Cars rush past through the dream-memories of patients. Freud looks on, forgetting he is prey to the primitive desires of the accelerating contemporary world, too—forgetting that 'science' is on occasion no less primitive in aspect than age-old myth.

In succumbing to the temptation to find in just one dream image the universal signification of his own psychoanalytic presence, Freud collapses the divide between the visual and conceptual which he has been at pains to establish. Even as he thinks of concept and thought as mastering the dream-image, decoding its concealed and shameful primitiveness, the speeding pictograph returns to invade and disrupt psychoanalysis itself. The spectacle of the fast vehicle is paradigmatic and yet also disruptive to the avowedly scientific spirit of Freud's modernizing rational discourse. The car is iconic of the modernizing industrialized putting-to-use of natural raw materials, of the plundering of nature (iron ore, rubber, oil, etc.) which exemplifies contemporary willing as the technical prowess that continually drives a speeding-up of the world. As image it is, by Freud's modern methodological standards, primitively symbolic, a mere visual icon awaiting proper interpretation at the hands of mastering thought. Instead of Freud mastering it, however, the speeding image masters him. For one thing, the image as exemplar of mechanical representation is already extant in the form of the photograph and primitive cinematic show which forms the basis of his model of the dream-theatre. Seduced by the temptation to worship speed, he rushes headlong to embrace this one form of dream image as being nothing less than a clear representation of himself and his theory. It is here, in this precipitate response to that dream image, rather than in the domain of supposedly 'primitive' incestuous drives, that there resides something genuinely archaic. It is the primitive appeal of strength and its promise of power over not just brute nature but others. In his response Freud shows himself to be prey to a tendency to respond uncritically to the visual promise of a technological prowess that will enclose the world in a safely regulative kingdom of reason. In an image of power he sees *himself* slicing through the landscape, masterful, bearing others helplessly along. Freud imagines he sees the power of his own speeding will and affirms it. He speeds past Oedipus.

SECTION 2: 'WHO OF US HERE IS OEDIPUS? WHO SPHINX?'[30]

The Sphinx, as is well known, poses a riddle.[31] Depending on the version of the tale, a wrong answer leads to the unfortunate respondent being smothered and devoured or flung from a cliff. Either way, you're toast. Oedipus answers correctly and the Sphinx obligingly responds by leaping to her own death. On Sophocles' representation of him, Oedipus holds this triumph over unruly nature to be wholly his own. It provides Oedipus with an empowering self-definition. He is foremost a figure who, at the outset of the play, sees himself as one who answers and solves the riddle by recourse to the power of his native cunning, his intellect.[32] His own quickness of wit, Oedipus boasts, is the source of his fame and power.[33] It is this which he thinks has installed him safely at the centre of enclosure of the Theban *polis* as its sovereign. As everyone knows, it turns out that the solution to the riddle is no prelude to resolution in a happily ever after. The solution encloses Oedipus. It marks the moment of his enthroned containment in the city as just another step toward his ultimate shame, for the city's enclosure is the individuating space which, by identifying Oedipus, contains him in a way that allows his subsequent public exposure as murderous and incestuous polluter.

Celebratory self-assertion is Oedipal, not some repressed desire to slice up the father because of fixation on the mother. Given that one's life is never concluded until death, such a celebration is also by its very nature precipitate. It is rightly said that no one ought to be called happy until he or she has 'crossed the border of life without enduring pain'.[34] Self-glorification and its leap to unwarranted conclusiveness is hubris—the pride that rises like a mist over the triumphant victories which mark our struggles with nature and history (which, so far as they relate to the latter, means with us ourselves). Hubris, to put it slightly differently, is the arrogance of 'confidence'. It is the faith that springs from feeling enclosed within a domain which, because it is closed in, seems amenable to our power of will and so capable of insulating us from the unexpected—the erroneous belief that the key to control of the future lies in absolute, radiating control over the present. This faith is still around and is not just the prerogative of powerful individuals. It is, for instance, the strange non-substance that helps lubricate contemporary liquefying global economies (international conditions of 'market confidence'). Economics is sovereign confidence addiction. It is the contemporary faith that action when it is the free play of autonomous human will in a market structure has a tendency to make things turn out for the best. What global stock market could function without 'confidence' in its own tomorrow, without faith in the resilience of human willing and in its ability to respond to resistances and overcome them? Hubris is, to put it another way, our unquestioning faith that through our dominant ways of doing things our future meaning is secured, that the meaning to which we now cleave will endure and in its enduring ensure our futurity.

Confidence is in this regard temporally orientated. Its court of judgement is the future. Economics and politics (which are conjoined in the currently dominant neo-liberal form of globalized existence) might in this regard be deemed the most recent of the great metaphysical edifices of the West in their Socratic-style promise of 'rational' transcendence. For what promise is not, insofar as it is about the future, essentially metaphysical in that it necessarily concerns what will be the case rather than what is? One can, of course, make an honest promise—for example, that 'the fool will stay'.[35] But this is not promising that things will turn out for the best, for that is never within anyone's power. Tragedy understands this. Tragedy understands that promising the future is disingenuous. Economic and political discourse, so long as it proffers the promise of material success, cannot therefore be mindful of tragedy since tragedy strips the promise back to its bare hope of constancy and undermines any politics of 'hope' that would trade on the future as if it were something that could be secured. When confidence gets blown away, like a mist, there is usually a crash . . . Tragedy or farce?

Insofar as Freud confidently conceives of himself and his method as the symbolic equivalents of industrialized technology, he, too, is 'Oedipal' in the above sense. His faith in the promise of powerful machinery as the epitome of progress leaps ahead of him. Culture frames the blindness of method. Freud has faith in his ability to solve the riddles of the mind by way of the cunning of intellect ensconced in the machinery of psychoanalytic tools. But when one makes an assertion of the order of Freud's, such as when he confidently judges that he sees the latent image of his therapeutic self hurtling toward him out of all dream fantasies of racing vehicles, a question mark steps up that places the assertion itself in question. The question mark arises because the matter of every individual auto-motive dream is thereby treated precipitately, i.e. as if the meaning of every one of them were already settled in advance according to an ordering that opposes primitive ignorance of symbolism to modern, rational, interpretative dexterity (a technical-scientific prowess that is the prerogative not just of modern machinery but analysis, too). Hubris is shorthand for the condition which accompanies this propensity to precipitate judgement. Hubris is the Sphinx of the question mark that inserts itself alongside confident Oedipal self-assertion.[36]

Oedipus and the Sphinx are entwined . . . hubris and fate. Again, therefore, 'Who of us here is Oedipus? Who Sphinx?' Do we ever only solve a riddle, as method promises? Or does our desire to see and celebrate in ourselves the source of its solving make riddles of us to ourselves? Nothing is ever simply a matter of asking a question and giving an answer which concludes a puzzle. The answer, as 'solution', does not simply contain the puzzle within itself. Answers are also puzzles that entail working through. They do not merely conclude things; they also begin them again. One needs here to listen to Nietzsche—which does not mean one needs to agree entirely with him. For him, the ancient Sophoclean entanglement of response and continuation of the riddle constitutes an abiding feature of the logic of modernity. Modernity, on the one hand, is hubris pure and simple.

> Even using the yardstick of the ancient Greeks, our whole modern existence is nothing but *hubris* and godlessness, in so far as it is strength and awareness of strength rather than weakness [. . .] *Hubris* today characterizes our whole

> attitude towards nature, our rape of nature with the help of machines and the completely unscrupulous inventiveness of technicians and engineers; *hubris* characterizes our attitude to God, or rather to some alleged spider of purpose and ethics lurking behind the great spider's web of causality.[37]

Modern life is nothing less than desire for control over nature by technical means and a rebellion against the notion of a pre-ordained order of things. Here is the death of the sacred. As such, modernity marks the era of the most comprehensive putting-to-use of the environment yet known. That said, as Nietzsche's description implies, modernity is for the most part devoid of self-reflexivity when it comes to understanding the full resonance of this feature which dominates it. The technician and engineer uncritically apply 'inventiveness' (that is, their own mental and practical dexterity) in a manner that absorbs them entirely in the problem at hand. They make use of things and concepts with little or no concern for the hidden aspects and implications of such use—the technician and engineer is the person who simply 'gets on with it' insofar as he or she unreflectively accedes to authority. One who accedes to authority pursues the purposes allotted them: in the midst of work one simply ponders how to extract an ore, bring about a chemical reaction, or build a structure more effectively and efficiently, etc., not why one is doing so or the order of command which legitimizes such activities. Modern life in this way is inclined to unthinking violation of the environment. Underneath such frenzied doing is the presupposition that the natural world is a passive object of human attention, a mere thing waiting to be taken hold of and subjugated by the authority of will.

Implicit in Nietzsche's description of modern life is activity on an industrial scale: an emergent moment of hypermodernization. We approach the natural environment with tools and look upon it as if it were a means to our satisfactions. We ruthlessly tear it to pieces with little regard for anything beyond our desires. We approach the earth with an eye to use—which means with an eye to its incorporation into ourselves.

Nietzsche is sensitive to the ambivalent forces at work here, for he poses an ambiguous ethical question in its wake. To the extent that we turn on nature we also turn on ourselves and become self-vivisectionists and self-violators. To question and challenge the environment necessarily entails questioning and challenging ourselves. This being so, is it not reasonable to ask whether we must in consequence 'become daily more deserving of being questioned, more *deserving* of asking questions, more deserving—of living?' Having posed the question, Nietzsche offers an ambiguous response, not an answer: 'All good things used to be bad things at one time; every original sin has turned into an original virtue'. Like the question, such a response pays testimony to the fact that the dominant order of things is changing in a manner that both compounds and renders problematic our link with the past. Industrial humankind turns on nature with tools and forces it to conform to its demands. Practices which once would have been condemned as evil are in the ascendant. Modernity, in line with Nietzsche's conception of the shifting changes in power which make of history a fragmented domain marked by power take-overs,[38] is entering the era of pure use, the age of economic justification. The standard of measure within such a culture is use itself (the end, the 'target'). Even reason comes to be revealed by modern sensibilities to be nothing more than a mere 'tool'.[39] The modern does not equate with anything more than an instrumental form of 'progress'—i.e. doing things 'better' (more efficiently) rather than living better. But this modern condition is also accompanied by feelings of ancient hubris: the past sense of a guilt that springs from rebellion against superior powers haunts it. Nietzsche does not so much seek to oppose as disrupt this state of affairs in his own perverse way by pointing to this archaic element. Pointing to the one he then invokes another. In the midst of invoking modern hubris—the collision of the ancient in the new—he also invokes something else archaic which modernity's technically inclined sensibility cannot easily cope with but which nevertheless abides within it as a companion of the hubris it unwittingly embraces. This is the question of modernity's sense of destiny, of what will become of it, which it raises in the very moment it seizes on nature and plunders it with industrial machinery. This is the matter of the *fate* of the modern.

Fate is what lies outside the bounds of all willing and technique. Contrary to Freud, it is not method-friendly. To love fate, Nietzsche tells his readers, is the 'formula for human greatness' for it means 'not wanting anything to be different, not forwards, not backwards, not for all eternity. Not just enduring what is necessary, still less concealing it [. . .] but *loving* it.'[40] With this, Nietzsche seeks to throw in the face of his audience something inimical to modern taste: he makes something that lies beyond the sphere of conventionally understood acts of willing and control the highest standard of measure. Fate does not come down to the things one wills to happen. It is the remainder, the excess which accrues within the hubristic dynamic of the assertion of will. It is what must happen to one whether one feels one has willed it or not. Fate indicates a place beyond redemption. The affirmation of fate entails affirming even one's failures, inevitable miseries, and ultimate perishing in the face of living. Loving fate means repudiating the idea that one can judge or vouchsafe anything beyond one's own sphere of life. It requires a relinquishment of the idea of a willing that can be separated from the domain out of which it emerges and on which it depends. Love of fate means affirming helplessness as not merely a necessary but central condition of being oneself.

> No one *gives* man his qualities: neither God, nor society, nor his parents and ancestors, nor man *himself* [. . .] *No one* is responsible for simply being there,

> or for being made in such and such a way, for existing under such conditions, in such surroundings. The fatality of one's being cannot be derived from the fatality of all that was and will be. *No one* is the result of his own intention, his own will, his own purpose; *no one* is part of an experiment to achieve an 'ideal person' [. . .] One is necessary, one is a piece of fate, one belongs to the whole, one *is* in the whole—there is nothing which could judge, measure, compare, condemn our Being, for that would mean judging, measuring, comparing, condemning the whole . . . *But there is nothing apart from the whole!*[41]

Nietzsche knows that in his times, which are modern times,[42] there can hardly be a more unpalatable gesture than urging the anti-theological embracing of a lack of control over the totality of things. Fate is what we who are becoming modernized (willingly or otherwise) would wish to seize with our own hands in order to bind it to a meaning that makes it implicitly our own. But fate is that which, in being reached for, eludes the modernizing liquefying grasp which would absorb everything within itself. No one can seek to absorb and so control the future without acknowledging that the attempt is likely at best to take its place alongside other delusions in the repetitive tragicomedy of human life.[43]

To love fate, to cleave to such love now of all times, must surely be amongst the hardest and most perverse of things. Modern fate, which is framed by modern cosmology, places us on a stone in a void which will ultimately melt away into what, from the standpoint of the living, must count as nothing. Nietzsche understands this too. An early mock-historical sketch imagines intelligent animals having sprung up near a warming star—before long the star cools and the clever creatures must die.[44] What this leaves for modern life to ponder is a fate that consists of environmental conditions that transcend it not by enclosing existence metaphysically but in the menacing thought of their material withdrawal. The creature that loves the idea of its own future above all things is, in the end, futureless.

To love fate, given this, means to stand within modernity as someone divided, as a being both at home in modern life and yet also set apart from and against it. To do that, Nietzsche comments in *Twilight of the Idols*, one must learn to see differently from the norm.[45] That is, one must learn *not to will*, i.e. not to respond straight away to every stimulus that rises up before one by reaching out toward it with curious and uncritically welcoming hands. To have to approach everything with all one's 'doors standing open' signifies the essentially modern taste for sweeping up and consuming everything 'new' one comes across, 'in short, our famous modern "objectivity"', the pseudo-science of 'value neutrality' which gobbles up everything under the illusion that everything must be potentially useful. Seeing differently means refusing to see in the way that is now the norm, which means seeing at a different tempo, for once one has begun to learn, 'as a learner you will have become generally slow, mistrustful, reluctant'. One will, in short, come to see in a manner that is resistant to precipitate judgement, to any hurried tendency to rush in, to the seduction of the power of acceleration which compels Freud to celebrate himself as its epitome. Not to hurry is a prerequisite of loving fate. Such love shatters the modern.

For Nietzsche, secular modernity is the cultural space of a dilemma in which an accelerating primitive desire for speed and archaic hubris collide. It is the environment in which slowness is condemned as outmoded, weak, inefficient, unproductive. This is because modern space is always already an Oedipal space.

SECTION 3: OEDIPUS FAST AND SLOW

Oedipal space: Oedipus is in a hurry. His city, Thebes, is 'wasting away'[46] from disease and he needs to find out why. The people suffer. In fact, Oedipus claims that, being king, he suffers even more than those who sicken and die. Where an individual citizen's grief is always private, his, he assures everyone, is public and therefore greater. The sovereign alone must bear the pain of the many.[47] Assuming this to be the case, it might seem reasonable that Oedipus wants to find the most efficient path to dealing with the crisis as quickly as possible. But it is also clear that this hurry is driven by an impatience that testifies to an inability to accept that sooner or later 'events will take their own course'.[48] Oedipus is resistant to the idea that things have a tendency to unfold independently of human willing. For him, the faster things can be made to happen the closer the harmony between them and him. Success in hurrying things along is taken as evidence of the power of will to determine the nature of things, even to the point of contradicting the will of the gods.

This conjoining of impatience and the urge to control events is made overt soon enough in *Oedipus Tyrannus*. A messenger returns with the oracle's answer as to why the city is mired in pestilence. The king's brother-in-law, Creon, advises that the news might be best heard in private, but no . . . Oedipus wants to know without delay. Hence, everyone hears. The problem is the 'pollution' which springs from the long-unsolved murder of the previous king of Thebes, Laius. The murder has not been atoned for.[49] This has brought about the displeasure of the gods for the killer lives on in the city. It is time for someone to pay. Now everybody knows.

The dimensions of Oedipus's impetuosity multiply. We hear that on his brother-in-law's advice he has already sent two men to get 'the gods' own seer',[50] Tiresias, who might be able to proffer an explanation for why things have gone bad. Oedipus cannot wait: 'I'm surprised he's not here already'.[51]

Not much later, Tiresias arrives to be barraged by impatient questioning. Oedipus, frustrated by the old man's silence (understandable, after all, since what he has to tell will tear Oedipus to pieces), threatens to rip the truth out of him.[52] Once Tiresias has made public the knowledge that Oedipus himself is the 'godless polluter of this land',[53] Oedipus rushes again, this time to repudiate his interlocutor. In a flurry, he accuses Tiresias and Creon of treasonable intent and asserts his power accordingly. If they want to usurp him, he will have them killed. The stately art of kingship thereby becomes permeated with haste also and is even justified by way of it when Creon objects to the unjust accusation: 'Still', Oedipus responds, 'I must rule'.[54] In other words, what makes sovereign power is the necessity of decision-making. It is not a matter of the 'right' decision, only of the timely need to decide, and meeting that need means reacting as rapidly as possible.

Evidence of urgency proliferates. There is Oedipus's precipitancy in imposing a sentence in advance on the murderer whose identity is yet unknown,[55] his willingness straight away to blame the people of Thebes for not attending to the matter for so long,[56] his rapid dismissal of Tiresias,[57] his demand that the only witness to the murder be quickly brought to him from the country.[58] There is his confession of how he fled in haste from Corinth having heard of the curse which he bears[59] and how, on his journey from there, he flew into the rage that sparked his killing of who would turn out to be his father, Laius.[60] Haste characterized the speedy manner of his despatching.[61] There is also Oedipus's hurrying of the messengers who ultimately confirm he is the criminal he seeks. Consider, above all, the acceleration into chaos which marks Oedipus's enraged self-mutilation when, having found his queen-wife-mother dead by her own hand, he puts out his eyes with her broaches in a frenzy of violence, proclaiming himself 'cursed by his own curse':[62]

> This was his chant as, again and again, not just once,
> He attacked and stabbed his eyes; and all the while
> His bleeding eyeballs spattered his chin,
> [Releasing no slow drops of gore, but all at once
> A dark shower, a hailstorm of soaking blood].[63]

Oedipus rushes in and everything rushes out of Oedipus. He is a kind of conduit for the regulation of flows which exceed and overwhelm him. Words, like his 'gore', gush out 'again and again'; even his bleeding is not a slow oozing but an explosion. He is the drive for speed-at-all-costs personified. In all he does, Oedipus races toward his own fate headlong, precipitating his predicament, the unwilling willer of his own destruction.

Sophocles reveals in this way a tension inherent in the relationship between willing and speed. On the one hand, Oedipus places faith in his ability to escape from the power of fate through flight. But the acceleration involved in flight turns inexorably back upon itself, initiating a resurgence and return of the very thing from which it is supposed to facilitate liberation. A sovereign confidence in the ability to out-riddle nature by speed of wit, the faith that the future can be controlled in such a way as to subvert what fate decrees by fleeing from it, secures Oedipus's downfall by facilitating his plunging headlong into it.

Sophocles conjoins this precipitate element with the realm of sight. Haste is blind. Confidence is blind. Few can fail to notice Sophocles' numerous and often ironic references to vision. With regard to the murder of his father, Oedipus describes himself as 'a stranger to what was done'[64] and the wanted killer as invisible—he knows a crime has been committed, 'but no one can see who did it'.[65] When Oedipus sees his father on the road he also does not recognize him for whom he is. Worse still, enthroned as king, Oedipus unknowingly comes to see through eyes that are rightfully those of another, namely, the murdered victim and father:

> I hold the authority he once had,
> I sleep and sow my seed where he once sowed his.
> My children would have shared a mother with his,
> If he had not the misfortune to be childless.[66]

Wilfulness is repaid with repulsive familiarity. In taking on Laius's role, Oedipus comes to see with Laius's eyes. He witnesses what he should not in his intimacy with his mother who now becomes also his children's mother. His vision, like everything else about him, is perverse. Even where he sleeps there is, in reality, no rest, for he is actively compounding the unwilled pollution which accompanies all his willed deeds.

So long as he has eyes, Oedipus cannot see who he is. His vision, in essence self-regarding, is invisible to itself, for it does not register the traces it leaves in its passing. There is what Oedipus sees and what, in casting his eyes about, his vision passes over. Oedipus does not stop to ponder this possibility or what it might imply, which is why first he by turns ridicules Tiresias for his sightlessness[67] and naivety and then contradictorily accuses him of being a devious 'money grabbing trickster with a good eye / For gain'. Another insult flips things over again almost immediately: 'Tell me', Oedipus crows, 'when have you proved clear-sighted?'[68] Prophecy, Oedipus now proudly contends, is contemptible. It is opposed to the virtue of sharp-sighted intelligence, which he personifies:

> How was it that when the singing bitch was here
> You said nothing to these citizens to set them free?
> And yet her riddle was not for any passer-by

To solve; it required a seer.
[. . .] Then I came along.
I, ignorant Oedipus, and I stopped her.
I used my wits; I didn't rely on birds.[69]

The light of reason is held to outstrip attempts to foretell the future by primitive recourse to spilling the guts of sacrificial fowl. Rational thought travels at the speed of mental agility which glories in its contrast to the slow vagaries of conventional wisdom, which is in this way held up in contempt for its impotence. How could Oedipus here not pour down his scorn thus since for him prophecy is fearful: he already knows it condemns him to something dreadful? Tiresias cannot but help allude to the irony of this in the aftermath of being accused of treason: 'Your eyes are shut / To your own—your very own, yet you find fault with me';[70] 'you cannot see the peril you are in'.[71] The blindness that encloses Oedipus is, in other words, threaded through the very fabric of his everyday being. His is a world dominated by his self-image as a person endowed with the autonomous power of willing capable of challenging the gods and nature and holding them at bay.

Oedipus feels he controls all he surveys, but he cannot see what surveys him; he does not notice that power in empowering also hems one in amidst the fatality of living with others. In Thebes, his authority is absolute, enclosed as he is within the defensive structure of public order which individuates him as king. Yet it is this which, by engendering the feeling of power, facilitates the further cultivation and strengthening of Oedipus's fatal lack of visual imagination, a lack rooted in the belief that the appearance of power *is* simply power.

Tiresias knows different. 'Ah, how terrible it is to be wise'.[72] One suspects Tiresias did not act against the Sphinx because this was not meant to be. Wisdom means understanding that nothing can be sped past or through with impunity, i.e. without something of what one is trying to avoid intermingling itself in the act of avoidance and binding itself to it. Nothing ever goes one way only. In seeing one is seen. In surveilling one is surveilled. In controlling one is controlled. In fleeing away from things and people, one flees into them. In concealment one is exposed. This is a condition of Apollonian individuated life, of the life lived within city walls. Oedipus had to come. By the same token, the light of Oedipus's wit has to be extinguished. The moment of dreadful self-awareness, of realizing he is who he is, makes Oedipus turn upon himself. Oedipus's eyes offend him, so immediately he puts them out. If Oedipus's speeding away from his fate is a kind of violence, this speeding returns to him as violence, slowing him down to a crippling gait. A slowed Oedipus is the riddle uttered by the Sphinx once more: a blind old man walking on three legs, the tapping of his stick signalling the evening of his life.

One might ponder the nature of this act of self-blinding. Is it Tiresian acceptance or Oedipal defiance? Is it both? On the one hand, blinded, Oedipus is helpless. He has therefore seemingly accepted his fate. Yet in blinding himself he nevertheless acts, asserting his will so as to frustrate what he sees awaiting him in the future. After all, a deceased, blinded Oedipus cannot look in shame on his parents when he arrives in Hades.[73] Likewise, in blinding himself he both (1) fulfils his own will by enacting the ban he in effect placed on himself as someone no longer fit to gaze on his own city[74] and (2) avoids the prospect of seeing his children having to face a life disgraced (his own eyes could not have desired such[75]). Oedipus tells us that he would have destroyed his hearing, too, would it have been possible. To be blind and deaf would be to be apart from all people: 'How sweet for the mind to dwell cut off from evil'.[76] Blind Oedipus is a symbol of the necessary conflict between human ways of doing things and the environment which such ways must negotiate. He is a symbol of what Heidegger deems the rip that scars the opening out of cultural space: of the necessary (and necessarily painful) gap that persists between any world and the environment of the earth.[77] Oedipus tries to close the gap, to bring the environment to heel—and he must fail. Monumental failure is a necessary condition of culture, for human living is necessarily caught up in what it cannot control as a condition of being there.

SECTION 4: SPEED OF TRAGEDY

This Oedipal flight remains at work in hypermodern culture. We, too, flee into the faith that our instrumental abilities will save us. Flight is distance and the greater the distance, the safer we feel. Greater distance is achieved by greatness of speed and greatness of speed is light speed. We turn the eye to serve the light of human wit to guide us into a future which is closed, to break open the unseen, to predict with technological cunning. But in such vision the hubristic rip which sustains culture opens up again. Thus Freud is outstripped and can be accused of hubris as soon as he dares to name the thing that triumphs over nature as his own—the assumption that psychoanalysis is symbolic of the hubristic ambition of the power of technology. This triumphant vision expresses a primitive Oedipal desire to make reality transparent to the manipulative light of an all-embracing human will. Its affirmation, however, returns Freud to the status of a primitive worshipper. In affirming the power which illuminates the dream worlds of his patients, he affirms Oedipal light, the light of willing which purports to answer the riddle with a promise capable of freeing us from servitude to nature by subordinating it to human will.

The Oedipal tale, in contrast, cultivates the paradox of will. For it suggests we learn a lesson that we cannot learn: that tragic fate always comes out of the blue, that the unexpected is to be expected even though it can never be anticipated. Fate is, to this extent, the mysterious, the unseen, that which exceeds calculation and all manner of visualization. Fate cannot be the object of sight since future disaster is always invisible. Its invisibility is what makes it inevitable and knowable only after the fact. Oedipus's faith, however, is our faith: the faith that, ultimately, nothing once it is rendered visible can really threaten us. A recoiling in fear of the dark. This is also the faith that manipulation will ultimately succeed in rendering everything sufficiently transparent and cooperative as to be pliant to human calculative willing. But there always remain in any calculation those aspects which exceed it. To have faith that making visible is equivalent to control is, for this reason, delusory. Making visible, for one thing, is not comprehensive; it is selective. All insight, as Paul de Man notes, comes at the cost of a necessary blindness.[78] Visuality is not 'objective' or 'value free'; it is engaged and judgemental; as well as all-encompassing in ambition, it is narrow with regard to what it must ignore in order to be what it is. One can have neither pure visibility nor pure darkness, only ambiguity and ambivalence (any image of catastrophe is in our eyes always tainted by an element of hope; every prospect of tragedy has the potential to proffer itself to someone's disinterested gaze as potential 'entertainment').

Sight requires light speed and is, due to this reliance, several steps behind that speed, which encompasses and outstrips it as the unseen condition of what can be seen. The speed of tragedy is the speed of light. Tragedy is the acknowledgement that disaster is the overwhelmingly accidental that must be experienced as a savage and fatal necessity because its rising to the level of visual awareness signals its unavoidably having already come to pass. Oedipus's fate precedes him precisely to the extent that he is condemned never to notice the moment of its impending arrival, which is a totality consisting of successive moments of his own arriving, e.g. at the place where three roads meet, the site at Thebes of the Sphinx, etc. He is destined to find himself trapped in an enclosing chaos which springs precisely from his accelerating attempts to impose an order on things in pursuit of escape from something inescapable. This is why, in the end, Oedipus can only simultaneously rave against and acknowledge his fate by depriving himself of the sight that was his hope of salvation.

Fate and will coexist. The one cannot be asserted without the other coming invisibly into play. This tension is a persistent and archaic condition of Western culture (let us not for the moment make it 'universal') for it patterns its self-understanding, driving the urge to attain a completed form of control. Speaking in the terms elaborated at the end of part 1 of this volume, Oedipus, even as king, is no straightforward Apollonian figure. Before his downfall he is that blend of Apollonian enclosure and Socratic ambition for transcendence which seeks to unite the realm of vision and the concept with an objective order amenable to human willing. Oedipus wants a world permeated by the light of human intelligence, one ordered by being made transparent and pliable to the demands of will. What explodes out of him in his acceleration toward his dreadful destiny is the luminescent Dionysian element which will not be represented in advance by visual strategies that seek to contain it and explodes precisely under the provocation of any attempt at containment. Apollonian, Dionysian, and Socratic visuality all come together to form the Oedipal tragedy.[79] The coexistence of fate and will pervades both the Apollonian defensive construction against Dionysian excess, which must be enclosed in order to be controlled, and the former's Socratic elaboration into a doctrine of pure knowledge that justifies an all-knowing sight of absolute knowledge. Socratism spurns the empirical and the body because these are vehicles of fate that must be tamed by will. But there is always more about a body and what may happen to it than one can see or know. Fate is invisible to the illuminating light of conceptual thought in that it cannot be predicted and calculated for. This is what makes it, like Dionysus, unreasonable: fate is 'immoral' insofar as 'doing the right thing' will make no difference. It will certainly not make us happy (as Socrates contended).

Even if the idealism of Socratism (its rejection of the empirical) seems redundant amidst the dominant materialism of contemporary culture, its spirit lives on in the contemporary ambivalent relationship with any mode of experience that cannot be exposed, regulated by method, and through that subjected to the taming and controlling power of methodical will. A Socratic visuality persists in globalized hypermodernizing form no less than it does in the ambivalent cultural space explored by Sophocles' engagement with the noble-contemptible figure of Oedipus. The contemporary love of exposure (of being seen, of one's life been recorded in every detail and transmitted to the wider world) and the suspicion of the hidden (the idea that what cannot be made transparent is a priori suspect or, worse, evil) are symptoms of this. Out of hatred of fate springs the desire for transcendence through manipulation. It is this desire that makes us, too, subject to the speed of tragedy.

SECTION 5: 'LUCKY AS A KNIFE'?

As a rule we seek to avert the tragedy of disaster by means of foresight in conjunction with the use of tools that allow us to manipulate the environment. We surround ourselves with the comforts of our technical means. Today, seeing ahead and equipment are more closely related to one another than ever before. They form a kind of material-perceptual laminate which strives to bind together cultural and environmental elements by coordinating them. In liquefying modern global space, tools for seeing and surveilling of all kinds now wind around the earth, enclosing it within a ring of communications and information flows which link points across its surface to ones in its upper atmosphere which in turn point back down. This is our continuation and intensification of the hubristic organization of practices that Nietzsche, as has been noted, identifies as quintessentially modern. We, alongside Oedipus and Socrates and Freud, are all believers in the age-old union of human will and vision. It is not a matter of freedom to decide. We are compelled by a desire that exceeds the individual to tie the earth up nice and safe by seeing it all over, all at once, and all the time.

It is not unreasonable to say that because of this the question of what might become of us today is bound up ever more closely with the ancient (non-Freudian) Oedipal relationship and its link to burgeoning modern spatial-visual technologies. This hypermodern binding together of faith in human will and technology as counters to fate is so intimate that the two permeate one another ever more relentlessly. Tools interlace in self-understanding, threading links in unexpected ways through heterogeneous realms of practices as and when they appear on the scene, transforming them, suggesting new meanings emerging from established practices that intimate the establishment of new regimes of power. Indeed, to the extent that technology gets interpolated into visuality, it constitutes the essential condition in virtue of which the thought of the future is today possible. We are led by our equipment, which in claiming to show us the future at the same time looks to us like the future. (Can almost anyone today imagine a *future without technology* that would be his or her own?) Above all, we are more and more led by the *technicity* of our equipment, that is, by its automaticity rather than any artisanal skill it might engender. Most of our equipment does what it does independently of us. We perhaps even derive much of our sense of willing from this automatic condition. One's feeling of 'willing' is allied to the sense of closeness which springs from the machine's automatic obedience to a pre-ordained command structure (the computer programme and what it relies on in order to function). The future does not belong to the quaint workbench of the artisan.

Given this last point, it is strange to consider that strictly speaking, tools of any kind can seem so often like our intimates such is our familiarity and closeness to them.[80] Yet they have no fate. To the extent this is the case, they exist both beside and beyond us. Animals, when they cease to be valued instrumentally, tend to face either extinction or absorption into other structures of equipmental exploitation, or they become cultural icons and means of 'entertainment' like the horse.[81] A tool is either put to use or abandoned to the historical dereliction of its outmodedness. Tools are destiny-less. Like the passing of the steam engine, the approaching near demise of the film camera or the telephone booth leaves in its wake a minimal historical sense of resonant nostalgia which calls upon the dim passing of older ways of life.[82] Those feelings aside, they are gone; they linger only as residual, spent waste products.

Fate is what clings to persons rather than things or technology. Even if the thing somehow becomes symbolic of a person or people, it is still a person or persons whose destiny resonates in it. It is still people, for now, who suffer misfortune. No person is 'lucky as a knife'.[83] Who could be so 'fortunate' as to be so much at one with their destiny as a piece of equipment is to its pre-destined use? This 'misfortune' is, of course, something that allows us to set ourselves apart from mere things and tools. One might dare to venture that, even if it were magically to become self-aware, equipment could not suffer even from misuse. A sentient knife could take neither joy nor misery from the cutting into bodies to which it was well or badly suited. The computer, however complex ('intelligent'), can never lament what it must do by design, for the tool is not uselessly embodied. It is hard, in short, to envisage how thought and the fate that clings to it could continue without a suffering body—that is, a body which suffers in virtue of being the kind of body it is, a body without determinate purpose and so fateful.[84] Without the possibility of useless suffering there is no lamentation, no tragedy, and therefore no need for human thought. Suffering begets confusion. Only a confused body is a body of culture and practices, a self.

What does this imply in a modern, hypermodernizing world where the intimacy of the relation between person and equipment is becoming more intense, closer? Might machines, as is the common fantasy, become more like us? Or will we become more akin to them and shape our ways of doing things according to their dominant kind of functionality? People already talk of 'the mind' as if it were a computer analogue—an ancient tendency to think metaphor in new clothes. One can already imagine the potential of this auto-technical intimacy which extends beyond that of the familiarity associated with the artisanal tool. It would most likely be an intensified variety of the intimate-impersonal relationship of the person who is glued by the cultural coordination of vision and movement to their mobile phone or computer (which of us now is not?) and through that is plugged into the wider environment. It is not hard to ponder a time when, due to technical facility, this relationship is enhanced: 'bespoke' equipmental services on a mass scale will be available. They are already here in the satellite box which monitors, reflects, and guides viewing habits; in the website which suggests news or music that may appeal on the basis of tracking the subject's relation to the electronic screen; in devices whose individual 'preferences' can be set; or in the promise of medical treatments that will cater for mapped genetic vulnerabilities. All these things observe, monitor, chart, and so enclose the self electronically, flattening it out into the plane of functional connectedness, hypermodernizing it. In doing so they constitute a radical reinterpretation of the Socratic dream of seeing life as a totality which can be enclosed and ordered by way of the light of an authoritative, calculative, all-seeing rationality that dissolves the remnants of outmoded tradition and subdues the unruly Dionysian body with the lure of higher disciplinary pleasures.

What, one wonders, might hypermodernizing culture achieve if it were to enclose suffering humanity within a technicity capable not merely of surrounding but also permeating it, making it at once malleable and impermeable to damage? This is certainly a dream of techno-enthusiasm. Would not what is promised be nothing less than an approach to the utopian heaven of ease and immunity from pain that equipment already apparently 'enjoys'? Lucky as a knife . . . beyond hurt? Technology has long been armour. Now

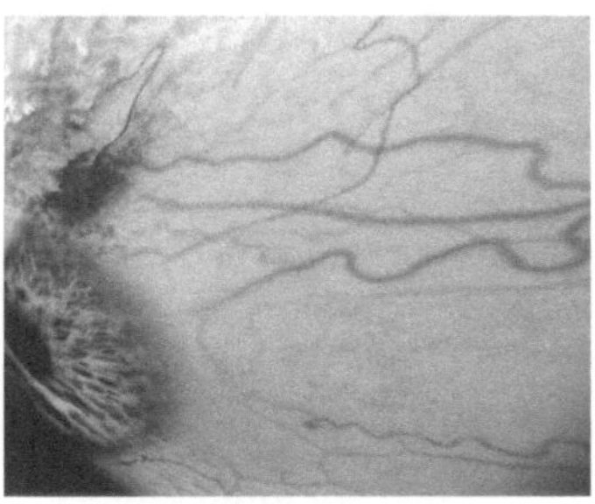

it is also endowed with more than the suggestion of some day becoming wholly at one with skins and organs. Prosthetic hands and eyes cannot suffer in touch and sight for they risk nothing of themselves. Just as the light of thought with transcendent ambition seeks to surround the earth with a plane of coordinates, replicating a military desire to conquer space by employing and refocusing surveilling practices, so the already socially enclosed Dionysian body, if it were to become entirely surrounded and saturated, too, might someday be faced with the prospect and lure of becoming transcendently immunized from pain. And this would surely require making of the earth a giant prosthesis—'humanizing' it entirely by making it pure equipment, unlimited enclosure, a gigantic synthetic eyeball whose battle-blooded scarring by historical existence would come in the end to amount to nothing more than dents on the hardened protective territorial exterior from out of which it looks down onto the plane it seeks to dominate.

A world entirely subject to the dream of such a willing would purport to be one that exists beyond pain, beyond fate. It would be a 'well managed' utopia which harnesses the primitiveness of the drives not by containing them but by suffusing the body itself with the sovereign technicity of method, permeating it entirely with the materially realized ancient demand of 'rational' regulative thought.

If culture is the general condition in which humanity comes about as the animal which seeks to enclose the unpredictable biology of the Dionysian earthly body by enclosing it within Apollonian layers of convention, then the selves who emerge from this channelling of body into culture revisit this unpredictability on each other in virtue of their unshakeable Dionysian inheritance. No Socratic reason, equipment, method, or strategic Oedipal calculation so far has overcome this condition since the rational, the equipmental, the methodological, and the calculative spring from and are bound to the unruly body they implicitly crave separation from. If a utopian hypermodernization were really to seek to free us from the bonds of fate by way of a technological perfection of the cravings of will, it would need to pursue the route of technology further than simply harmonizing the self with technology. Liberation from fate would require the eradication of the antinomy of unpredictable others altogether from the field of life's encounters. The environment would have to be wrapped entirely in a network of transparent layers that converted everything into surveilled modes of connected and regulated visibility. The sky itself would become an enclosure . . .

A world would be required in which encounters were no longer face to face but always mediated between devices—truly predictable: like a video game, only more 'real'. This would be seeing only by way of the enclosure of a frame that proclaims its right to encompass everything. The screen would need to be triumphant in its distancing of person from person, in its transcendent enactment of personhood's withdrawal into the machine's enclosure . . .

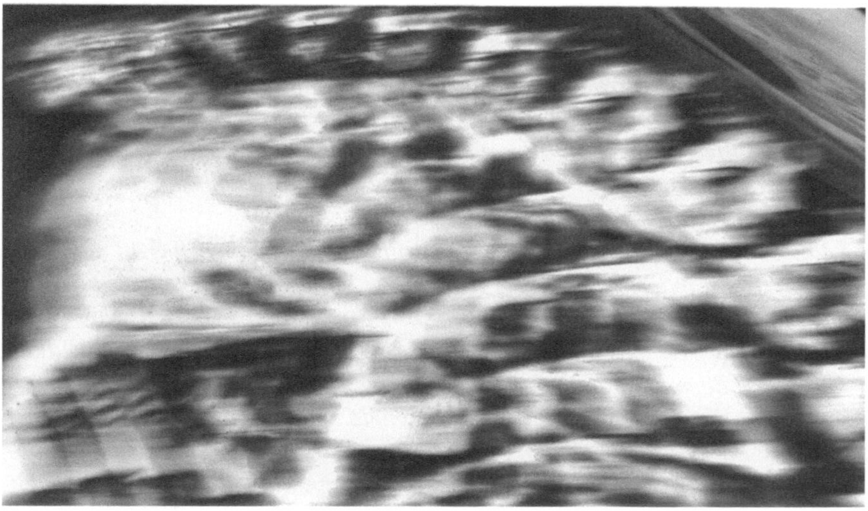

X-Ray

When the surgeon showed you
how your lungs were shot,
all you saw were Rorschach blots

until the creature came –

man-hydra, nine-faced and indivisible,
rippling in time's pool,
its bodies ribbed like sand,
powerless to face itself.

A machinic enclosure. Slices of movement are spread across a flattened plane by a strobe light. Here is a body cut up. Light here hints at its capacity to be turned into a scalpel in conjunction with lens and sensitive surface. This conjunction dissects time and its relation to bodies in motion, catching isolated and flitting moments and making them appear concrete with pulsing alternations of light and dark. The dark recedes. It forms a background. The light comes forward, throwing the subject into relief. Look away and then look back. Seeing the image again, one starts with what seems obvious. One is drawn to a sequence of faces. These faces form a blurred arc that falls or rises across the plane of the image's frame (left to right, right to left: one can see it either way). The features are anonymous. Yet, here is conveyed an unconscious intimacy—one intimated by mechanical means. Each face relates intimately to the other because each, although different and blurred and mask-like, is clearly the same. We are confronted here with a question of time. Each face is revealed as what becomes or has become of the other. Each rests upon or supports the other in its capacity as a face caught in a movement both away from and toward itself. We understand ourselves as who we might become, but this becoming is invisible to us. The image is, in this regard, a little snapshot of fate—or chance, which after the fact becomes fate and before the fact is the last desperate hope of those who feel they may be condemned by fate. Who is not thus condemned?

A closer look at the faces reveals a strong contrast between light and dark. This contrast renders eye sockets as emptied out spaces. The figure might be blind. Even the sighted, however, are also blind in an essential way. In the midst of sight we remain blind to the future for we cannot see it. Yet we have faith in and depend upon the future's coming to light for there is no meaning in darkness. Faith: even in darkness we would rather will than not will at all. Does this claim imply a kind of self-blinding as a condition of human action? That willing as it is often conceived (as the illumination of the world by intention) is an illusion? If so, then human action, however casual, would have to be characterized foremost by blind striving. Blind striving in turn testifies to a mad hope in relation to fate and chance. This is the hubristic hope that things can be overcome by a dominant power of vision, that 'light' will triumph. This hope springs from a persistence of tradition, from its language of light and dark. We who see now are all, in this regard, present in this image. Fate has long been a matter of darkness erupting in light, of blindness and insight after the fact emerging from the present to disrupt hubristic pretensions. What might it mean to want to transcend this condition? This, too, is an old, old question with an old still-living answer. Above the figure to the right floats a gigantic eye. It is incongruous . . . coarse . . . transcendent . . . unfeeling . . .

NOTES

1. David Kovacs, 'The Role of Apollo in *Oedipus Tyrannus*', in *The Play of Texts and Fragments: Essays in Honour of Martin Cropp*, ed. J. R. C. Cousland and James R. Hume (Leiden: Brill, 2009), 357–58.

2. For this view, see E. M. Harris, 'Is Oedipus Guilty? Sophocles and Athenian Homicide Law', in *Law and Drama in Ancient Greece*, ed. E. M. Harris et al. (London: Duckworth, 2010), 122–46.

3. See L. Versényi, 'Oedipus, Tragedy and Self Knowledge', *Arion* 1, no. 3 (1962): 20–30.

4. See Philip Vellacott, *Sophocles and Oedipus: A Study of* Oedipus Tyrannus *with a New Translation* (London: Macmillan, 1971). It becomes apparent to the '*careful reader*' (104) that Oedipus's nearly two-decade-long care for the city is an attempt at atonement for a wrong he knows he has committed (119). When Apollo demands the debt be paid off, Oedipus initiates an investigation, hoping for the facts of the matter

to come out in a way that somehow relieves him of the cruel task of telling his citizens the truth himself. This rather unconvincing argument is framed within a discussion of the play as charting the logical uncovering of the truth of the awareness Oedipus already has concerning his origins.

5. Kovacs, 'The Role of Apollo in *Oedipus Tyrannus*', 368.

6. See Bernard Knox, 'Introduction to *Oedipus the King*', in Sophocles, *The Three Theban Plays*, trans. Robert Fagles (London: Penguin, 1984), 145.

7. To refresh the memory, here is a summary. Against good advice, King Laius and Queen Jocasta of Thebes have a son, Oedipus. The oracle of Apollo at Delphi foretells that the child of their union will murder his father and marry his mother. Given the fact that desire has trumped good sense, it is decided that the only thing to do is to have the baby killed. A servant is commanded to leave baby Oedipus to die on the side of a mountain. Just to make sure he dies, Oedipus's feet are pinned together. (Oedipus is named after the scars this pinning causes; his name means 'Swollen Foot'.) Killing a child, however, is not something the servant can conscience doing. He gives the child to a shepherd he knows in the mountains who, it turns out, is the servant of another king. This king, Polybus of Corinth, adopts Oedipus and makes him his heir. When grown to a young man, Oedipus hears the prophecy concerning him from a divine messenger. He decides to frustrate the prophesy and he leaves Corinth and the man he thinks to be his father. On a road approaching the city of Thebes, Oedipus is drawn into a violent altercation with a group of men. He kills them all. A powerful being known as the Sphinx holds Thebes in its fearful sway. It torments the city, posing a riddle no one seems capable of answering. Oedipus arrives. He solves the riddle. The Sphinx is destroyed. Thebes celebrates. However, nobody currently occupies the throne of the city. This is because the previous incumbent has been killed on a road by a stranger. Oedipus, defeater of the Sphinx, is made the new king of Thebes and takes the widow of the previous king to be his wife and queen. Many years of prosperous life pass until suddenly plague and death come to the city. An oracle reveals that the god Apollo has commanded this suffering as a consequence of the failure to bring the previous king's murderer to justice. The play begins with this time of pestilence. Oedipus orders an investigation. He moves things on as fast as possible. As past events come to light, it becomes evident that the murderer who is sought by Oedipus is Oedipus. Oedipus is revealed to have unwittingly killed his own father Laius and to have taken as his wife his own mother. The four offspring of the union are hence both Oedipus's children and his siblings at one and the same time. Jocasta's response to the revelation is suicide; Oedipus's response is to blind himself horrifically with her broach pins. Disgraced and blind, Oedipus must face the sentence of banishment he himself commanded as punishment for the murder.

8. Freud's classic study of this phenomenon is the case of Little Hans. See Sigmund Freud, 'Analysis of a Phobia in a Five-Year-Old Boy', in *The Standard Edition of the Complete Works of Sigmund Freud*, trans. and ed. James Strachey (New York: Vintage, 2001), 5–148.

9. The claim is made by Strachey in his introduction to Sigmund Freud, *The Interpretation of Dreams*, trans. and ed. Lytton Strachey (New York: Basic Books, 2010), xviii (italics mine).

10. Freud, *The Interpretation of Dreams*, 278.
11. Freud, *The Interpretation of Dreams*, 279.
12. Freud, *The Interpretation of Dreams*, 280.
13. Freud, *The Interpretation of Dreams*, 280.
14. Freud, *The Interpretation of Dreams*, 604.
15. Freud, *The Interpretation of Dreams*, 280.
16. Gilles Deleuze and Felix Guattari, *Anti-Oedipus: Capitalism and Schizophrenia*, trans. Robert Hurley, Mark Seem, and Helen R. Lane (Minneapolis: University of Minnesota Press, 2000), 50.
17. Deleuze and Guattari, *Anti-Oedipus*, 23.
18. Deleuze and Guattari, *Anti-Oedipus*, 108.
19. Freud, *The Interpretation of Dreams*, 295.
20. Freud, *The Interpretation of Dreams*, 295–96.
21. See Friedrich Nietzsche, *On the Genealogy of Morality*, ed. Keith Ansell Pearson, trans. Carol Diethe (Cambridge: Cambridge University Press, 2017), 7, in which Nietzsche talks of 'decoding the hieroglyphic-script of the human moral-past'.
22. A plague of vermin signifies pregnancy; an air-ship (a rather recent symbol), 'the male organ'; etc. (Freud, *The Interpretation of Dreams*, 371).
23. Freud, *The Interpretation of Dreams*, 419. The comment is dated 1919.
24. A connection in Freud between the 'inextricably linked' worlds of 'the new automobiles [. . .] the only slightly older railroad journey' and imaging technology is noted in passing by Friedrich Kittler in the context of their alienating effects (*Optical Media*, trans. Anthony Enns (Cambridge: Polity 2010), 169). Freud, on a railway journey, is disconcerted one day to discover that the sight of an old man 'reflected in the mirrored door of a first-class bathroom' 'whose appearance "he thoroughly disliked"' is 'his own mirror image' (181). For some discussion of these matters, see Thomas Elsaesser, 'Freud and Technical Media: The Enduring Magic of the Wunderblock', in *Media Archaeology: Approaches, Applications and Implications*, ed. Erkki Huhtamo and Jussi Parikka (Berkeley: University of California Press, 2011), 95–115.
25. See Sigmund Freud, 'Analysis: Terminable and Interminable', in *The Standard Edition of the Complete Works of Sigmund Freud*, vol. 6, trans. and ed. James Strachey (London: Hogarth Press, 1964), 216–53. Analysis, says Freud, is 'a time consuming business' (216). One of the central concerns of the piece is how the whole analytic process might be speeded up.
26. Freud, *The Interpretation of Dreams*, 408n1.
27. Freud, *The Interpretation of Dreams*, 408.
28. Freud, *The Interpretation of Dreams*, 280.
29. Freud, *The Interpretation of Dreams*, 281.
30. Friedrich Nietzsche, *Beyond Good and Evil*, in *Basic Writings of Nietzsche*, trans. Walter Kaufmann (New York: Basic Books, 1968), sec. 1.
31. The riddle was 'What goes on four legs in the morning, two in the afternoon, and three at night?' The answer Oedipus gave was 'Man'.
32. For an excellent study of this in relation to the work of Adorno and Horkheimer on the cunning of instrumental rationality, see Christopher Rocco, *Tragedy and Enlightenment* (Berkeley: University of California Press, 1997), chap. 2.

33. 'Famous Oedipus, as everyone calls me'. Sophocles, *Oedipus Tyrannus*, trans. Ian McAuslan and Judith Affleck (Cambridge: Cambridge University Press, 2003), 25, 390ff. All references to the text are by line number.

34. Sophocles, *Oedipus Tyrannus*, 1530.

35. William Shakespeare, *King Lear*, ed. R.A. Foakes (Oxford: Arden Shakespeare, 2005), act 2, scene 4.

36. It is why there are no 'indubitable truths' that can present themselves as objects for rational assertion. To assert is always already to presuppose.

37. Nietzsche, *On the Genealogy of Morality*, III, sec. 9.

38. See Nietzsche, *On the Genealogy of Morality*, II, sec. 13–14, discussed in part one of this volume.

39. See Nietzsche, *Beyond Good and Evil*, 191: 'Reason is only an instrument'.

40. Friedrich Nietzsche, *Ecce Homo*, trans. Duncan Large (Oxford: Oxford University Press, 2007), sec. 10.

41. Friedrich Nietzsche, *Twilight of the Idols*, trans. Duncan Large (Oxford: Oxford University Press, 1998), 8.

42. For an excellent exposition of this, see Laurence Lampert, *Nietzsche and Modern Times: A Study of Bacon, Descartes and Nietzsche* (New Haven, CT: Yale University Press, 1995).

43. See the discussion offered in Friedrich Nietzsche, *The Gay Science*, trans. Walter Kaufmann (New York: Random House, 1974), sec. 1. The history of culture which Nietzsche envisages here is tragicomic. The notion that life has a unifying meaning is identified here as the source of tragedy. Every religious and moral teaching in history has promised an absolute meaning but suffered its inevitable decline and fall, usually at the hands of some fatal accident of chance. With this, the brief time of 'tragedy always gave way again and returned to the eternal comedy of existence'. This marks the essence of the human dilemma. For although all teachings of ultimate purposes are doomed from the outset to destruction in the face of a totality whose contingencies they cannot accommodate, their combined historical consequence has been fundamental in shaping human identity. Religious and moral illusions have preserved the species and with that the notion that existence has a purpose and meaning has become a 'need' no less important to humankind than food and drink. Nietzsche here picks up on the implications of Darwin's theories. The autochthonous requirement for salutary meaning is part of 'the amazing economy of the preservation of the species' and has made humankind a 'fantastical animal' that '*has to* believe, to know, from time to time, *why* he exists' and to have 'faith in reason in life' in order to continue to exist. This need for meaning makes human history possible. What matters is the need and not the supposed 'content' (meaning) of any particular teaching. This need also ensures that humankind is doomed to the eternal repetition of the same errors.

44. Friedrich Nietzsche, 'On Truth and Lies in a Non-moral Sense', in *The Nietzsche Reader*, ed. Keith Ansell Pearson and Duncan Large (Oxford: Blackwell, 2006), 114. This brief period is 'world history'.

45. Nietzsche, *Twilight of the Idols*, sec. 6.

46. Sophocles, *Oedipus Tyrannus*, 25.

47. Sophocles, *Oedipus Tyrannus*, 62–64.

48. Sophocles, *Oedipus Tyrannus*, 341.
49. 'Something must be exiled or a death must pay for a death' so as 'to make those who killed him [Laius] pay' (Sophocles, *Oedipus Tyrannus*, 96, 100, 107).
50. Sophocles, *Oedipus Tyrannus*, 299.
51. Sophocles, *Oedipus Tyrannus*, 290.
52. Christopher Rocco comments that 'the seer's prophetic, riddling, and symbolic knowledge stands in direct opposition to the king's analytical rigor, coherent logic, coherent method, and highly structured and structuring intellect' (*Tragedy and Enlightenment*, 43). Haste, however, is not analytical or rigorous, it is regulative, methodological, proto-scientific in ethos—which in why, as Rocco also notes, Tiresias stands as a living 'rebuke to the mastering intellect of Oedipus' (44).
53. Sophocles, *Oedipus Tyrannus*, 353.
54. Sophocles, *Oedipus Tyrannus*, 627.
55. Sophocles, *Oedipus Tyrannus*, 216ff.
56. Sophocles, *Oedipus Tyrannus*, 256.
57. Sophocles, *Oedipus Tyrannus*, 445.
58. 'Could we get him back here again—and quickly?' (Sophocles, *Oedipus Tyrannus*, 765).
59. Sophocles, *Oedipus Tyrannus*, 787ff. Oedipus is told that he will sleep with his mother and kill his father. Since he believes his parents to be the king and queen of Corinth, he flees from them.
60. Sophocles, *Oedipus Tyrannus*, 805ff. These others being Oedipus's father, Laius, and servants.
61. 'For quickly I hit him with the staff in my hand [. . .] I killed them, every one' (Sophocles, *Oedipus Tyrannus*, 810–13).
62. Sophocles, *Oedipus Tyrannus*, 1292.
63. Sophocles, *Oedipus Tyrannus*, 1275–79.
64. Sophocles, *Oedipus Tyrannus*, 220.
65. Sophocles, *Oedipus Tyrannus*, 293.
66. Sophocles, *Oedipus Tyrannus*, 259–62.
67. Oedipus: 'You're blind—ears, mind, eyes—all blind [. . .] You're wrapped in unending night; you'll never harm / Me or anyone else who sees the light of day' (Sophocles, *Oedipus Tyrannus*, 371–75).
68. Sophocles, *Oedipus Tyrannus*, 387–90.
69. Sophocles, *Oedipus Tyrannus*, 396–98.
70. Sophocles, *Oedipus Tyrannus*, 337–38.
71. Sophocles, *Oedipus Tyrannus*, 367.
72. Sophocles, *Oedipus Tyrannus*, 316.
73. Sophocles, *Oedipus Tyrannus*, 1371–73.
74. Sophocles, *Oedipus Tyrannus*, 1381.
75. Sophocles, *Oedipus Tyrannus*, 1388.
76. Sophocles, *Oedipus Tyrannus*, 1388–90.
77. Heidegger, 'The Origin of the Work of Art', in *Basic Writings*, ed. David Farrell Krell (London: Routledge, 1978). Also see part 2 chapter 2 of this volume.

78. See Paul de Man, *Blindness and Insight: Essays in the Rhetoric of Contemporary Criticism* (London: Methuen, 1983). De Man's primary interest in this text is the work of the New Critics who came to prominence in literary studies in the 1950s. The New Critics were committed to the idea that any literary piece of writing is an organic unity. However, de Man shows, this commitment is contradicted by the critics' own practice, which in its pursuit of unearthing the organic elements in literature deploys metaphors which undercut the idea itself. Often, the New Critics' metaphors of organic unity compromise the very notion that they are supposed to exemplify by introducing ambiguities of meaning which come to disrupt it. The unearthing of such ambiguities is, for de Man, the New Critics' signal achievement. Yet this achievement is bought at the cost of their blindness when it comes to the nature of their own critical language.

79. The reading proffered here might be said to run counter to Christopher Rocco's contention that 'Sophoclean wisdom [. . .] prefigures itself as something of a Socratic enigma', anticipating Socrates' '"structured deconstruction" of the social order and of itself as part of that order' (*Tragedy and Enlightenment*, 65). Rocco regards the structure of Plato's *Republic* as being amenable to being read in terms that are less totalizing and 'monolithic' than might be suspected and no less fractured by 'internal tensions, ambiguities and paradoxes than the tragedy it rejects' (114).

80. This intimacy, as Kittler notes, is brought to the fore with the invention of high definition television, which brings with it the promise of enveloping the viewer, transporting them from the 'simulated intimacy' of the conventional television screen to a relation of 'total intimacy' with the spectacle that thereby absorbs them (*Optical Media*, trans. Anthony Enns (Cambridge: Polity 2010), 223).

81. For an informative account of our fading relationship with horses, see Ulrich Raulff, *Farewell to the Horse: The Final Century of Our Relationship*, trans. Ruth Ahmedzai Kemp (London: Allen Lane, 2017).

82. A good example is offered by Michael Pritchard, *A History of Photography in 50 Cameras* (London: Bloomsbury, 2014), which charts the history of photographic practice through the development of its equipment. The narrative begins with the first production camera, Alphonse Giroux's daguerreotype (1839), and concludes with the Nokia Lumia 1020 smartphone (2013).

83. Robyn Hitchcock, 'Luckiness', from *Propeller Time* (Sartorial Records, 2010). Luck is what makes us who we are and determines who we will become—our faces, those whom we happen to meet or marry, our preferences and propensities. Luck, the song notes, may be a guide, like a dog that steers us blind toward the future. The future, in other words, is darkness, unknowable, that which dwells beyond the domain of the seen and becomes visible only as it erupts in the now and then is already gone. To be lucky means to be destined. 'Lucky as a knife': a knife cuts through things with ease. Its destiny, unlike a person's, is harmoniously bound up with its use function. Nobody is a knife or even wholly like one, one suspects. No one is ever simply lucky. That is human destiny.

84. See Jean-François Lyotard, 'Can Thought Go On Without a Body?' in *The Inhuman*, trans. Geoffrey Bennington and Rachel Bowlby (Oxford: Polity, 1991), 8–23.

Lyotard questions whether artificial intelligence could ever encompass the conditions of thought which mark human intellectuality. Can intelligence, in other words, be separated from the mortal body, which is 'the only available *analogon* for thinking a certain complexity of thought' (22)? Such complexity, for Lyotard, lies at the heart of human inventiveness, whose 'analogizing power' (23) does not conduct itself by way of alternating 0s and 1s (i.e. in computer language) but turns on the 'phenomenological body' and the 'irreparable transcendence inscribed on the body by gender difference' (23). This matter is linked to suffering most explicitly by Nietzsche (see *On the Genealogy of Morality*, III, sec. 13). Humankind, Nietzsche argues, is '*the* sick animal' precisely because it suffers from itself, that is, from the question of what will become of it. Human futurity, of course, is fate.

Appendix: Conversation

- That picture makes me sad.

◦ I like it. It takes me back to a moment that's gone. The handprints suggest someone is between places. The door may be closed, but the latch is open.

- What do you mean, 'between places'? They are in that country from which we are all exiled: childhood. And the door is 'closed, but the latch is open' because you never mended it. That's why the shed rotted and we had to pull it down.

◦ You mean: I had to pull it down . . . 'Between places' is not opaque. It's where any sense of future is. 'Between' is promise, on the way to somewhere from somewhere.

• That reminds me of *leurad* in Welsh: the piece of land which is the first to dry when the tide goes out which is neither the seabed nor the dry land but an 'in between'. Not one thing or another. A magical place, temporary, a boundary and yet part of the two sides it divides. The handprints are like that, in a way.

◦ You mean, they are a child's but we know they are also now almost an adult's? I said I pulled down the shed, but actually the maker of those prints helped me . . .

• Yes. That's why the photograph of his handprints makes me sad because he is no longer that child. Perhaps it reminds me of the boundary I'm crossing. Was it mud on his hands? One set of prints looks muddy and the other, paler.

◦ That day it rained and rained. Then there was late sun. I remember the afternoon's low sun shedding light on the shed's surface. A pool of mud came from a kind of war dance, I remember. The prints are pure mud.

• It's got the orange tint of clay, the stuff of birth and death. It's a cave painting and a piece of graffiti.

◦ A cave man and an urban criminal!

• I wonder which is the more civilized. Those ochre paintings of bison and birds or spray-painted bubble writing?

◦ The images in the caves of Lascaux are said to be 17,000 years old . . .

• The handprints look very happy, like they're waving to us. A hello or a goodbye.

◦ Or both. As I said, they are between places. There is something of play in these hands. Some say that many of the hand outlines in Neolithic caves were made by children's hands.

• It's funny how people have always made prints of their hands. Why do we love our hands so much?

◦ If you look at cave paintings you realise that there's something about the hands and the eyes which goes together. Sight and touch go together. The image has to be made using the hand and you reach out to touch the image even in 'just looking'. The painter in the cave, one feels, touches the animal with sight. In our culture the ancient relationship between sight and touch has been dissolved because we mostly produce images with machines. But the ancient occult quality of the image remains.

- That's very interesting, but I think people make handprints because our hands are who we are. They represent our choices because we have two hands: 'this or that'. They are our actions and our morals. Everything we do we do with our hands.

◦ Ah yes, good old morality! The eye as an extension of the hand. Seeing is tied to doing. Everything we do comes down to the conversations which can occur between the hand and the eye.

Bibliography

Adorno, Theodor W. 'The Essay as Form', trans. Bob Hullnot-Kentor and Frederic Will, *New German Critique* 32 (Spring–Summer 1984): 151–71.

Agamben, Giorgio. *Homo Sacer: Sovereign Power and Bare Life*, trans. Daniel Heller-Roazen (Stanford, CA: Stanford University Press, 1998).

———. *Stasis: Civil War as a Political Paradigm*, trans. Nicholas Heron (Edinburgh: Edinburgh University Press, 2015).

Andermatt Conley, Verena. 'The Passenger: Paul Virilio and Feminism', *Theory, Culture and Society* 16, nos. 5–6 (1999): 201–14.

Armitage, John, and Ryan Bishop, eds. *Virilio and Visual Culture* (Edinburgh: Edinburgh University Press, 2013).

Ayer, A. J. *Language, Truth and Logic* (London: Victor Gollancz, 1967).

Bakhtin, Mokhail. *Rabelais and His World*, trans. Hélène Iswolsky (Bloomington: Indiana University Press, 2009).

Barbero, Alessandro. *The Battle: A New History of Waterloo*, trans. John Cullen (London: Atlantic Books, 2005).

Barney, Darin, Gabriella Coleman, Christine Ross, Jonathan Sterne, and Tamar Tembeck, eds. *The Participatory Condition in the Digital Age* (Minneapolis: University of Minnesota Press, 2016).

Barthes, Roland. *Camera Lucida*, trans. Richard Howard (London: Vintage, 2000).

———. *The Grain of the Voice*, trans. Linda Coverdale (New York: Hill & Wang, 1985).

———. *Image, Music, Text: Essays*, trans. Stephen Heath (London: Fontana, 1977).

Bartram, Rob. 'Visuality, Dromology and Time-Compression: Paul Virilio's New Occularcentrism', *Time and Society* 13, nos. 2–3 (2004): 285–300.

Baudelaire, Charles. 'The Painter of Modern Life', in *Selected Writings on Art and Literature*, trans. P. E. Charvet (London: Penguin, 2006).

Baudrillard, Jean. *The Transparency of Evil: Essays on Extreme Phenomena*, trans. James Benedict (London: Verso, 2009).

Bauman, Zygmunt. *Liquid Times* (Cambridge: Polity, 2007).

———. *A Chronicle of Crisis: 2011–2016* (London: Social Europe Edition, 2017).
———. *Retrotopia* (Cambridge: Polity, 2017).
Beals, Kurt. 'Text and the City: George Grosz, Neue Jugend, and the Political Power of the Popular Media', *Dada/Surrealism* 19, no. 1 (2013).
Beck, Harry. London Underground Map, 1933, https://tfl.gov.uk/corporate/about-tfl/culture-and-heritage/art-and-design/harry-becks-tube-map.
Beckman, Karen, and Liliane Weissberg, eds. *On Writing with Photography* (Minneapolis: University of Minnesota Press, 2013).
Benjamin, Walter. *The Arcades Project*, trans. Howard Eiland and Kevin McLaughlin (London: Belknap Press, 1999).
———. *Illuminations*, trans. Jonathan Zorn (London: Pimlico, 1999).
Bentham, Jeremy. *The Panopticon Writings*, ed. Miran Božovič (London: Verso, 1995).
Berger, John. *About Looking* (London: Bloomsbury, 2009).
———. *Understanding a Photograph* (London: Penguin, 2013).
———. *Ways of Seeing* (London: Penguin, 2008).
Birtwistle, Harrison, and Tony Harrison. *Yan, Tan, Tethera: A Mechanical Pastoral*, www.universaledition.com/themes/Frontend/UniversalEdition/frontend/_public/src/js/pdfjs/web/viewer.html?file=/resources/ansichtspartituren/ue17684_pa_a3.pdf.
Blake, William. *Songs of Innocence and Experience*, in *The Portable Blake*, ed. Alfred Kazin (Harmondsworth: Penguin, 1979).
Blakely, Edward J., and Mary Gail Snyder. *Fortress America: Gated Communities in the United States* (Washington, DC: Brookings Institution Press, 1999).
Boemke, Manfred F., Roger Chickering, and Stig Förster, eds. *Anticipating Total War: The German and American Experiences, 1871–1914* (Cambridge: Cambridge University Press, 1999).
Boym, Svetlana. *The Future of Nostalgia* (New York: Basic Books, 2001).
Bradley, Arthur, and Louis Armand, eds. *Technicity* (Prague: Litteraria Pragensia, 2006).
Bubić, Andreja, Ana Sušac, and Marijan Palmović. 'Observing Individuals Viewing Art: The Effects of Titles on Viewers' Eye-Movement Profiles', *Empirical Studies of the Arts* 35, no. 2 (2017): 194–213.
Bueno, Claudio Celis. *The Attention Economy: Labour, Time and Power in Cognitive Capitalism* (London: Rowman & Littlefield, 2017).
Burgard, Peter J., ed. *Nietzsche and the Feminine* (Charlottesville: University of Virginia Press, 1994).
Burke, Jason. 'How Can Cities Protect Their Citizens Against Vehicle Attacks?' *Guardian*, 10 August 2018.
Butler, Richard. '"Perfect" Sensors Might Be Possible But Might Not Come to Cameras', *Digital Photography Review*, 28 February 2018, www.dpreview.com/interviews/9385721576/perfect-sensors-may-be-possible-but-might-not-come-to-cameras.
Cadava, Eduardo. *Words of Light: Theses on the Photography of History* (Princeton, NJ: Princeton University Press, 1997).

Chickering, Roger, and Stig Förster, eds. *Great War, Total War: Combat and Mobilization on the Western Front, 1914–1918* (Cambridge: Cambridge University Press, 2000).

———, eds. *The Shadows of Total War: Europe, East Asia, and the United States, 1919–1939* (Cambridge: Cambridge University Press 2003);

Chickering, Roger, Stig Förster, and Bernd Grenier, eds. *A World at Total War: Global Conflict and the Politics of Destruction, 1937–1945* (Cambridge: Cambridge University Press 2005).

Christodolou, Daisy. *Seven Myths About Education* (London: Routledge, 2014).

Chul-Han, Byung. *The Transparency Society*, trans. Erik Butler (Stanford, CA: Stanford University Press, 2015).

Clark, Maudemarie. *Nietzsche on Truth and Philosophy* (Cambridge: Cambridge University Press, 1990).

Connerton, Paul. *How Modernity Forgets* (Cambridge: Cambridge University Press, 2009).

Crary, Jonathan. *Suspensions of Perception* (Cambridge, MA: MIT Press, 2001).

———. *Techniques of the Observer: On Vision and Modernity in the Nineteenth Century* (Cambridge, MA: MIT Press, 1992).

———. *24/7* (London: Verso, 2014).

Crogan, Patrick. 'The Tendency, the Accident and the Untimely: Paul Virilio's Engagement with the Future', *Theory, Culture and Society* 16, nos. 5–6 (1999): 161–76.

Csikszentmihalyi, Mihaly, and Mick E. Robinson. *The Art of Seeing: An Interpretation of the Aesthetic Encounter* (Malibu, CA: J. Paul Getty Museum and the Getty Center for Education in the Arts, 1990).

Cubitt, Sean. 'Virilio and the New Media', *Theory, Culture and Society* 16, nos. 5–6 (1999): 127–43.

Danto, Arthur C. 'The Artworld', *Journal of Philosophy* 61, no. 19 (1964): 571–84.

Darwin, Charles. *On the Origin of Species*, ed. J. W. Burrow (Harmondsworth: Penguin, 1976).

Davies, Bryan Martin. *Cerddi Bryan Martin Davies*, ed. Alan Llwyd (Llandybïe: Cyhoeddiadau Barddas, 2003).

De Man, Paul. *Blindness and Insight: Essays in the Rhetoric of Contemporary Criticism* (London: Methuen, 1983).

De Quincey, Thomas. 'The English Mail-Coach', in *Confessions of an English Opium-Eater and Other Writings*, ed. Robert Morrison (Oxford: Oxford University Press, 2013).

Deleuze, Gilles. *Empiricism and Subjectivity: An Essay on Hume's Theory of Human Nature*, trans. Constantin V. Boundas (New York: Columbia University Press, 1991).

———. *The Logic of Sense*, trans. Mark Lester and Charles Stivale (London: Athlone Press, 1990).

———. *Negotiations, 1972–1990*, trans. Martin Joughin (New York: Columbia University Press, 1995).

———. 'Postscript on Control Societies', in *Negotiations, 1972–1990*, trans. Martin Joughin (New York: Colombia University Press, 1995), 177–82.

Deleuze, Gilles, and Félix Guattari. *Anti-Oedipus: Capitalism and Schizophrenia*, trans. Robert Hurley, Mark Seem, and Helen R. Lane (Minneapolis: University of Minnesota Press, 2000).

Der Derian, James. 'The Conceptual Cosmology of Paul Virilio', *Theory, Culture and Society* 16, nos. 5–6 (1999).

Derrida, Jacques. *Margins of Philosophy*, trans. Alan Bass (London: Harvester Wheatsheaf, 1982).

———. *The Truth in Painting*, trans. Geoff Bennington and Ian McLeod (Chicago: University of Chicago Press, 1987).

Dewey, John. *Art and Experience* (New York: Perigree, 2009).

———. *The Influence of Darwin on Philosophy and Other Essays* (New York: Prometheus Books, 1997).

———. 'The Unit of Behavior (The Reflex Arc Concept in Psychology)', in *Pragmatism: The Classic Writings*, ed. H. S. Thayer (Minneapolis, MN: Hackett, 1982).

Duncan, Carol. *Civilizing Rituals: Inside Public Art Museums* (London: Routledge, 1995).

Eisenhower, Dwight D. 'The Military-Industrial Complex', in *21 Speeches That Changed Our World*, ed. Chris Abbot (London: Ryder/Random House, 2012), 243–44.

Flusser, Vilém. *Towards a Philosophy of Photography*, trans. Anthony Mathews (London: Reaktion, 2006).

Foucault, Michel. *Discipline and Punish*, trans. Alan Sheridan (New York: Vintage, 1977).

Fox, Rachel. 'Against Progress: Understanding and Resisting the Temporality of Transformational Weight-Loss Narratives', *Fat Studies* 7, no. 2 (2018): 216–26.

Freud, Sigmund. 'Analysis of a Phobia in a Five-Year-Old Boy', in *The Standard Edition of the Complete Works of Sigmund Freud*, trans. and ed. James Strachey (New York: Vintage, 2001).

———. 'Analysis: Terminable and Interminable', in *The Standard Edition of the Complete Works of Sigmund Freud*, vol. 6, trans. and ed. James Strachey (London: Hogarth Press, 1964).

Friedberg, Anne. 'Virilio's Screen: The Work of Metaphor in the Age of Technological Convergence', *Journal of Visual* Culture 3, no. 2:183–93.

Gane, Mike. 'Paul Virilio's Bunker Theorizing', *Theory, Culture and Society* 16, nos. 5–6 (1999): 85–102.

Gates, David. *The Napoleonic Wars, 1803–15* (London: Pimlico, 2003).

Gernsheim, Alison, and Helmut Gernsheim. *A Concise History of Photography* (London: Thames & Hudson, 1971).

Gilbert, Martin. *The First World War* (London: HarperCollins, 1995).

Gillespie, Michael Allen, and Tracy B. Strong, eds. *Nietzsche's New Seas* (Chicago: University of Chicago Press, 1988).

Greb, Fabian, Paul Elvers, and Timo Fischer. 'Trends in Empirical Aesthetics: A Review of the *Journal of Empirical Studies of the Arts* from 1983 to 2014', *Empirical Studies of the Arts* 35, no. 1:3–26, 18.

Grosz, George. 'Kannst du radfahren?' *Neue Jugend* (June 1917): 1.

Habermas, Jürgen. *The Future of Human Nature*, trans. Max Pensky, Hella Beister, and William Rehg (Cambridge: Polity, 2003).

———. *The Philosophical Discourse of Modernity*, trans. Frederick Lawrence (Cambridge: Polity Press, 1987).

Hacking, Juliet, ed. *Photography: The Whole Story* (London: Thames & Hudson, 2014).

Hansen, Mark. *Embodying Technesis: Technology Beyond Writing* (Ann Arbor: University of Michigan Press, 2000).

Hardy, Thomas. *Jude the Obscure*, ed. Patricia Ingham (Oxford: Oxford University Press, 2002).

Harris, E. M. 'Is Oedipus Guilty? Sophocles and Athenian Homicide Law', in *Law and Drama in Ancient Greece*, ed. E. M. Harris et al. (London: Duckworth, 2010), 122–46.

Harvey, David. *The Condition of Postmodernity* (Oxford: Blackwell, 1992).

Hayman, Stephen. 'Photos, Photos Everywhere', *New York Times*, 29 July 2015, www.nytimes.com/2015/07/23/arts/international/photos-photos-everywhere.html.

Heidegger, Martin. *Basic Writings*, ed. David Farrell Krell (London: Routledge, 1978).

———. *Being and Time*, trans. John Macquarrie and Edward Robinson (Oxford: Blackwell, 2005).

———. *Off the Beaten Track*, trans. Julian Young and Kenneth Haynes (Cambridge: Cambridge University Press, 2002).

Hitchcock, Robyn. 'Luckiness', *Propeller Time* (Sartorial Records, 2010).

Hobbes, Thomas. *Leviathan*, ed. Richard Tuck (Cambridge: Cambridge University Press, 1994).

Hobsbawm, Eric. *Age of Extremes: The Short Twentieth Century* (London: Abacus, 1995).

Horkheimer, Max. *Eclipse of Reason* (New York: Continuum, 1974).

Horkheimer, Max, and Theodor Adorno. *Dialectic of Enlightenment*, ed. Gunzelin Schmid Noerr, trans. Edmund Jephcott (Stanford, CA: Stanford University Press, 2002).

Huhtamo, Erkki, and Jussi Parikka, eds. *Media Archaeology: Approaches, Applications and Implications* (Berkeley: University of California Press, 2011).

Hume, David. *An Enquiry Concerning Human Understanding*, in *Enquiries Concerning Human Understanding and Concerning the Principles of Morals*, ed. L. A. Selby-Bigge (Oxford: Clarendon Press, 1988).

———. *A Treatise of Human Nature*, ed. L. A. Selby-Bigge (Oxford: Clarendon Press, 1990).

Huntington, Samuel P. *The Clash of Civilizations and the Remaking of the World Order* (London: Simon & Schuster, 1997).

———. *Who Are We? The Challenges to America's National Identity* (New York: Simon & Schuster, 2005).
Inwood, Brad, ed. *The Cambridge Companion to the Stoics* (Cambridge: Cambridge University Press, 2003).
James, Ian. 'Phenomenology in Diaspora: Paul Virilio and the Question of Technology', *French Cultural Studies* 17, no. 3 (2006): 319–33.
Jay, Martin. *Downcast Eyes: The Denigration of Vision in Twentieth-Century French Thought* (Berkeley: University of California Press, 1994).
Kadet, Anne. 'Metro-Money: Napping in a City That Never Sleeps', *Wall Street Journal*, 8 August 2018.
Kaufmann, Walter. *Nietzsche: Philosophy, Psychologist, Antichrist* (Princeton, NJ: Princeton University Press, 1974).
Kellner, Douglas. 'Virilio, War and Technology: Some Critical Reflections', *Theory, Culture and Society* 16, nos. 5–6 (1999): 103–25.
Kelsey, Robin. *Photography and the Art of Chance* (Cambridge, MA: Belknap Press, 2015).
Kittler, Friedrich. *Gramophone, Film, Typewriter*, trans. Goeffrey Winthrop-Young and Michael Wutz (Stanford, CA: Stanford University Press, 1999).
———. *Optical Media*, trans. Anthony Enns (Cambridge: Polity 2010).
———. *The Truth of the Technological World: Essays on the Genealogy of Presence*, trans. Erik Butler (Stanford, CA: Stanford University Press, 2013).
Knox, Bernard. 'Introduction', in Sophocles, *The Three Theban Plays*, trans. Robert Fagles (London: Penguin, 1984), 129–55.
Kovacs, David. 'The Role of Apollo in *Oedipus Tyrannus*', in *The Play of Texts and Fragments: Essays in Honour of Martin Cropp*, ed. J. R. C. Cousland and James R. Hume (Leiden: Brill, 2009), 357–68.
Kracauer, Siegfried. *The Mass Ornament: Weimar Essays*, trans. and ed. Thomas Levin (Cambridge, MA: Harvard University Press, 1995).
———. *Theory of Film: The Redemption of Physical* Reality (New York: Oxford University Press, 1960).
Kristeva, Julia. *Powers of Horror: An Essay on Abjection*, trans. Leon S. Roudiez (New York: Columbia University Press, 1984).
Kroker, Arthur. *The Possessed Individual* (London: Macmillan, 1992).
Kurgan, Laura. *Close Up at a Distance: Mapping, Technology, and Politics* (New York: Zone Books, 2013).
Lacquer, Walter, ed. *Fascism: A Reader's Guide* (Harmondsworth: Penguin, 1979).
Lampert, Laurence. *Nietzsche and Modern Times: A Study of Bacon, Descartes and Nietzsche* (New Haven, CT: Yale University Press, 1995).
Leach, Neil. 'Virilio and Architecture', *Theory, Culture and Society* 16, nos. 5–6 (1999): 71–84.
Levin, Bernard. *Conducted Tour* (London: Sceptre, 1988).
Lindell, Annukka K. 'Motor Biases Do Not Influence Posing Orientation in Selfies', *Laterality* 22 (2017): 49–59.
———. 'No Cheek Bias: Orientation in Van Gogh's Portraits and Self-Portraits', *Empirical Studies of the Arts* 35, no. 2 (2017): 127–38.

Lipovetsky, Gilles. *Hypermodern Times*, trans. Andrew Brown (Cambridge: Polity, 2005).

Locke, John. *An Essay Concerning Human Understanding*, ed. P. H. Nidditch (Oxford: Oxford University Press, 1998).

Loeb, Paul S. *The Death of Nietzsche's Zarathustra* (Cambridge: Cambridge University Press, 2010).

Lowry, Bernard. *British Home Defences, 1940–45* (Oxford: Osprey, 2004).

Lynch, Gordon. *The Sacred in the Modern World: A Cultural Sociological Approach* (Oxford: Oxford University Press, 2012).

Lyotard, Jean-François. 'Can Thought Go On Without a Body? in *The Inhuman*, trans. Geoffrey Bennington and Rachel Bowlby (Oxford: Polity, 1991), 8–23.

———. *The Differend: Phrases in Dispute*, trans. Georges Van Den Abeele (Manchester: Manchester University Press, 1988).

———. *The Postmodern Condition: A Report on Knowledge*, trans. Geoff Bennington and Brian Massumi (Manchester: Manchester University Press, 1984).

Lyotard, Jean-François, and Jean-Loup Thébaud. *Just Gaming*, trans. Wlad Godzich (Manchester: Manchester University Press, 1985).

Machiavelli, Niccolò. *Machiavelli: Selected Political Writings*, trans. and ed. D. Wootton (Indianapolis, IN: Hackett, 1994).

Magnus, Bernd, and Kathleen M. Higgins, eds. *The Cambridge Companion to Nietzsche* (Cambridge: Cambridge University Press, 1996).

Martindale, Colin. 'Recent Trends in the Psychological Study of Aesthetics, Creativity and the Arts', *Empirical Studies of the Arts* 25 no. 2 (2007): 121–41.

Marx, Karl. *Capital: A New Abridgement*, ed. David McLellan, trans. Samuel Moore and Edward Aveling (Oxford: Oxford University Press, 1995).

McLuhan, Marshall. *Understanding Media: The Extensions of Man* (Cambridge, MA: MIT Press, 1994).

McQuire, Scott. 'Blinded by the (Speed) of Light', *Theory, Culture and Society* 16, nos. 5–6 (1999): 143–59.

Miller, Jonathan. Channel 4 News, 25 May 2017.

Mitchell, William J. *The Reconfigured Eye: Visual Truth in the Post-Photographic Era* (Cambridge, MA: MIT Press, 1992).

Moholy-Nagy, László. *The New Vision and an Abstract of an Artist* (New York: Witternborn Schulz, 1947).

Müller, Christopher John, and Günther Anders. *Prometheanism: Technology, Digital Culture and Human Obsolescence* (London: Rowman & Littlefield, 2016).

Müller, Christopher John, and Ian Wiblin. 'The Bank of England in Ruins: Photography, Money and the Law of Equivalence', in *Credo Credit Crisis: Speculations on Faith and Money*, ed. Laurent Milesi, Christopher John Müller, and Aidan Tynan (London: Rowman & Littlefield, 2017), 87–127.

———. *Prometheanism: Technology, Digital Culture and Human Obsolescence* (London: Rowman & Littlefield, 2016).

Mulvey, Laura. 'Visual Pleasure and Narrative Cinema', in *Film Theory and Criticism*, ed. Leo Braudy and Marshall Cohen (Oxford: Oxford: University Press, 1999).

Mulvey, Laura, and Roberta Sassatelli. 'Interview with Laura Mulvey: Gender, Gaze and Technology in Film Culture', *Theory, Culture and Society* 28, no. 5:123–43.

New York Metropolitan Museum of Art. *360° Project*, www.metmuseum.org/art/online-features/met-360-project.

Nietzsche, Friedrich. *Beyond Good and Evil*, in *Basic Writings of Nietzsche*, trans. Walter Kaufmann (New York: Basic Books, 1968).

———. *The Birth of Tragedy*, in *Basic Writings of Nietzsche*, trans. Walter Kaufmann (New York: Basic Books, 1968).

———. *Daybreak*, trans. R. J. Hollingdale (Cambridge: Cambridge University Press, 1982).

———. *Ecce Homo*, trans. Duncan Large (Oxford: Oxford University Press, 2007).

———. *The Gay Science*, trans. Walter Kaufmann (New York: Random House, 1974).

———. *Human, All Too Human*, trans. R. J. Hollingdale (Cambridge: Cambridge University Press, 1982).

———. *On the Genealogy of Morality*, ed. Keith Ansell Pearson, trans. Carol Diethe (Cambridge: Cambridge University Press, 2017).

———. 'On Truth and Lies in a Non-moral Sense', in *The Nietzsche Reader*, ed. Keith Ansell Pearson and Duncan Large (Oxford: Blackwell, 2006).

———. 'On the Uses and Disadvantages of History for Life', in *Untimely Meditations*, trans. R. J. Hollingdale (Cambridge: Cambridge University Press, 1984).

———. *Thus Spoke Zarathustra*, in *The Portable Nietzsche*, trans. Walter Kaufmann (London: Chatto & Windus, 1971).

———. *Twilight of the Idols*, trans. Duncan Large (Oxford: Oxford University Press, 1998).

Noonan, Jeff. 'Paul Virilio and the Temporal Conditions of Philosophical Thinking', *Time and Society* (4 April 2017): 1–20.

Oakeshott, Michael. *On Human Conduct* (Oxford: Clarendon Press, 1991).

Osborne, Mike. *Defending Britain: Twentieth-Century Military Structures in the Landscape* (Stroud: History Press, 2010).

Peterson, Andrea. 'Everything You Need to Know About the Undersea Cables that Power Your Internet and Why They're at Risk of Breaking', *Independent*, 27 October 2015.

Plato. *The Laws*, trans. Trevor J. Saunders (Harmondsworth: Penguin, 1970).

———. *Republic*, trans. Desmond Lee (London: Penguin, 2007).

Pritchard, Michael. *A History of Photography in 50 Cameras* (London: Bloomsbury, 2014).

Raulff, Ulrich. *Farewell to the Horse: The Final Century of Our Relationship*, trans. Ruth Ahmedzai Kemp (London: Allen Lane, 2017).

Reid, Julian. 'Rethinking Relations between War, Modernity and City Spaces after 9/11', *Space and Culture* 7, no. 4 (November 2004): 396–408.

Rocco, Christopher. *Tragedy and Enlightenment* (Berkeley: University of California Press, 1997).

Rorty, Richard. *Philosophy and the Mirror of Nature* (Princeton, NJ: Princeton University Press, 1979).

Rose, Gillian. *Visual Methodologies* (London: Sage Publications, 2016).

Rosen, Stanley. *The Question of Being: A Reversal of Heidegger* (New Haven, CT: Yale University Press, 1993).

Safranski, Rüdiger. *Nietzsche: A Philosophical Biography*, trans. Shelly Frisch (London: Granta, 2002).

Schmitt, Carl. *The Concept of the Political*, trans. George Schwab (Chicago: University of Chicago Press, 1996).

Schumpeter, Joseph A. *Capitalism, Socialism, and Democracy* (London: Routledge, 2010).

Sedgwick, Peter R. 'Hobbes, Sovereign Power and Money', in *Credo Credit Crisis: Speculations on Faith and Money*, ed. Laurent Milesi, Aidan Tynan, and Christopher Müller (London: Rowman & Littlefield, 2017).

———. 'Hyperbolic Naturalism: Nietzsche, Ethics and Sovereign Power', *Journal of Nietzsche Studies* 47, no. 1 (Spring 2016): 141–66.

———. 'Nietzsche, Illness and the Body's Quest for Narrative', *Health Care Analysis* 21, no. 4 (2013): 306–22.

———. *Nietzsche's Economy: Modernity, Normativity and Futurity* (Basingstoke: Palgrave, 2007).

Shakespeare, William. *King Lear*, ed. R. A. Foakes (Oxford: Arden Shakespeare, 2005).

Sharer, Robert J., and Loa P. Traxler. *The Ancient Maya* (Palo Alto, CA: Stanford University Press, 2005).

Shawcross, Nancy M. *Roland Barthes on Photography: The Critical Tradition in Perspective* (Gainesville: University Press of Florida, 1997).

Silverman, Kaja. *The Miracle of Analogy* (Stanford, CA.: Stanford University Press, 2015).

Simmel, Georg. 'The Metropolis and Mental Life', in *The Sociology of Georg Simmel*, ed. Kurt H. Wolff, trans. H. H. Gerth and C. Wright Mills (Glencoe, Illinois: Free Press, 1950).

Smith, Adam. *The Wealth of Nations*, ed. Andrew Skinner (London: Penguin, 1999).

Smith, Jeffrey K., and Lisa K. Smith. 'Spending Time on Art', *Empirical Studies of the Arts* 19, no. 2 (2001): 229–36.

Sontag, Susan. *On Photography* (Harmondsworth: Penguin, 1977).

Sophocles. *Oedipus Tyrannus*, trans. Ian McAuslan and Judith Affleck (Cambridge: Cambridge University Press, 2003).

Soyinka, Wole. *Myth, Literature and the African World* (Cambridge: Cambridge University Press, 1979).

Stiegler, Bernard. *States of Shock: Stupidity and Knowledge in the Twenty-First Century*, trans. Daniel Ross (Oxford: Polity, 2014).

———. *Technics and Time, 1: The Fault of Epithemus*, trans. Richard Beardsworth and George Collins (Stanford, CA: Stanford University Press, 1998).

Strachey, James. Introduction to *The Interpretation of Dreams* by Sigmund Freud, trans. and ed. Lytton Strachey (New York: Basic Books, 2010).

Strauss, Leo. *Natural Right and History* (Chicago: Chicago University Press, 1965).

———. 'The Three Waves of Modernity', in *Political Philosophy: Six Essays by Leo Strauss*, ed. Hilail Gilden (Indianapolis, IN: Bobbs-Merrill, 1975), 81–98.

Stummer, Robin. 'Back to the Darkroom: Young Fans Reject Digital to Revive Classic Film Camera', *Guardian*, 28 January 2018.

Sturken, Marita, and Lisa Cartwright. *Practices of Looking: An Introduction to Visual Culture* (Oxford: Oxford University Press, 2001).

Taylor, Charles. *Philosophical Arguments* (Cambridge, MA: Harvard University Press, 1997).

TeleGeography.com. www.telegeography.com.

Thiong'o, Ngũgĩ wa. *Writers and Politics* (London: Heinemann, 1990).

Thucydides. *The Peloponnesian War*, trans. Martin Hammond (Oxford: Oxford University Press, 2009).

Tiggemann, Marika, and Isabella Barbato. '"You look great!"—The Effect of Viewing Appearance-Related Instagram Comments on Women's Body-Image', *Body Image* 27 (2018).

United Kingdom Home Office. *Air Raid Precautions: Directions for the Erection and Sinking of the Galvanised Corrugated Steel Shelter* (London: Home Office, 1939).

United States Library of Congress Rare Book and Special Collections Division. www.loc.gov/pictures/item/2006690453/.

Vellacott, Philip. *Sophocles and Oedipus: A Study of Oedipus Tyrannus with a New Translation* (London: Macmillan, 1971).

Venn, Couze. *Occidentalism: Modernity and Subjectivity* (London: Sage Publications, 2000).

Versényi, L. 'Oedipus, Tragedy and Self Knowledge', *Arion* 1, no. 3 (1962): 20–30.

Virilio, Paul. *The Art of the Motor*, trans. Julie Rose (Minneapolis: University of Minnesota Press, 1995).

———. *Bunker Archaeology*, trans. George Collins (Princeton, NJ: Princeton University Press, 1994).

———. 'The Illusions of Zero Time', trans. Chris Turner, in *Virilio and Visual Culture*, ed. John Armitage and Ryan Bishop (Edinburgh: Edinburgh University Press, 2013), 28–36.

———. *Open Sky*, trans. Julie Rose (London: Verso, 2008).

———. *Politics of the Very Worst*, ed. Sylvère Lotringer, trans. Michael Cavaliere (New York: Semiotext(e), 1999).

———. *Speed and Politics: An Essay on Dromology*, trans. Mark Polizzotti (Los Angeles: MIT Press, 2007).

———. *The Vision Machine*, trans. Julie Rose (Bloomington: Indiana University Press, 1994).

Virilio, Paul, and John Armitage. 'From Modernism to Hypermodernism and Beyond: Interview with John Armitage, *Culture, Theory and Society* 16, nos. 5–6 (1999).

Virilio, Paul, and Sylvère Lotringer. *Pure War*, trans. Mark Polizzotti (Los Angeles: Semiotext(e), 2008).

Von Clausewitz, Carl. *On War*, trans. O. J. Matthijs Jolles, in *The Book of War* (New York: Modern Library, 2000).

Vulliamy, Elsa. 'Nikon Photography Competition: Winning Entry Turns Out to Be Badly Doctored Image', *Independent*, 1 February 2016.

Weigel, Sigrid. 'The Flash of Knowledge and the Temporality of Images: Walter Benjamin's Image-Based Epistemology and Its Precondition in Visual Arts and Media History', *Critical Inquiry* 41, no. 2 (2015): 344–66.

Weller, Paul. *The Rushdie Affair* (London: Continuum, 2009).

Wilkes, Maggie, and Barry Gunter. *The Media and Bodily Image* (London: Sage Publications, 2004).

Wills, David. *Dorsality* (Minneapolis: University of Minnesota Press, 2008).

———. 'Raw War: Technotropological Effects of a Divided Front', *Oxford Literary Review* 31, no. 2 (2009): 133–52.

Wilson, Robert. *Great Exhibitions: The World Fairs 1851–1937* (London: Studio Vista, 1977).

Index

Index prepared by Nick Viner

About the Author and Poet

Peter R. Sedgwick is Reader in the School of English, Communication and Philosophy at Cardiff University. He is editor of *Nietzsche: A Critical Reader* (Blackwell, 1995). He is, with Andrew Edgar, co-editor of *Cultural Theory: The Key Concepts* (Routledge, 2nd revised edition 2007), and co-author of *Cultural Theory: The Key Thinkers* (Routledge, 2001). He is the author of *Descartes to Derrida* (Blackwell, 2001), *Nietzsche's Economy* (Palgrave, 2007), *Nietzsche: The Key Concepts* (Routledge, 2009) and *Nietzsche's Justice* (McGill-Queen's University Press, 2013).

Damian Walford Davies is the author of the poetry collections *Docklands* (2019), *Judas* (2015), *Witch* (2012) and *Suit of Lights* (2009)—all published by Seren—together with the libretto, *The Mare's Tale* (Grey Mare Press, 2019) and the pamphlet, *Alabaster Girls* (Rack Press, 2015). He is Pro Vice-Chancellor (Arts, Humanities and Social Sciences) at Cardiff University.

www.ingramcontent.com/pod-product-compliance
Ingram Content Group UK Ltd.
Pitfield, Milton Keynes, MK11 3LW, UK
UKHW040909190926
475560UK00008B/68

9 781786 604903